Genes, Brains, and Politics

Recent Titles in
Human Evolution, Behavior, and Intelligence

Staying Human in the Organization: Our Biological Heritage and the Workplace
J. Gary Bernhard and Kalman Glantz

The End of the Empty Organism: Neurobiology and the Sciences of Human Action
Elliott White

GENES, BRAINS, AND POLITICS

Self-Selection and Social Life

Elliott White

HUMAN EVOLUTION, BEHAVIOR, AND INTELLIGENCE
SEYMOUR ITZKOFF, SERIES EDITOR

PRAEGER
Westport, Connecticut
London

Library of Congress Cataloging-in-Publication Data

White, Elliott.
Genes, brains, and politics : self-selection and social life / Elliott White.
p. cm.—(Human evolution, behavior, and intelligence, ISSN 1063-2158)
Includes bibliographical references and index.
ISBN 0-275-94468-9 (alk. paper)
1. Neurobiology—Social aspects. 2. Genetics—Social aspects.
I. Title. II. Series.
QP356.W444 1993
304.5—dc20 92-28548

British Library Cataloguing in Publication Data is available.

Library of Congress Catalog Card Number: 92-28548
ISBN: 0-275-94468-9
ISSN: 1063-2158

First published in 1993

Praeger Publishers, 88 Post Road West, Westport, CT 06881
An imprint of Greenwood Publishing Group, Inc.

Printed in the United States of America

The paper used in this book complies with the Permanent Paper Standard issued by the National Information Standards Organization (Z39.48-1984).

10 9 8 7 6 5 4 3 2 1

To my wife Arleen

Contents

Tables and Figures

TABLES

FIGURES

Preface

I think of this book as a sequel to my earlier *The End of the Empty Organism*. Whereas that work was a critique of contemporary behavioral science, this volume delineates a positive alternative approach to the study of social and political action.

In this endeavor, as in the last, I have benefitted from my colleagues within politics and the life sciences, including Joseph Losco, Roger Masters, Glendon Schubert, and the late Thomas Weigele.

Earlier versions of chapters 6 and 7 (''Gene-Environment Interaction: The Neuropolitics of Alienation'' and ''Self-Selection and Sexual Politics'') appeared as ''Self-Selection and Political Life,'' *Politics and the Life Sciences* (February 1989) and ''Self-Selection and Sexual Politics,'' *Social Science Information* (September 1991).

I am especially indebted to Gloria Basmajian for her indispensable clerical services. When I counted on her, Gloria, as before, came ever reliably to my rescue.

The dedication, finally, is only a small indication of the appreciation due my wife Arleen for her years of patient support and encouragement.

Chapter 1

Introduction

Can *zoon politicon*, the political animal, be understood by leaving out both the "political" and the "animal"? That is precisely what contemporary behavioral science often does. Implicitly, if not explicitly, it commonly operates on premises that are environmentalist, mechanistic, and typological. The first rules out consideration of the "animal" and the second rules out "politics." The third, or typological, minimizes both the "political" and the "animal."

Before offering a counter-model, an IMPish paradigm—we shall briefly elaborate this critique. A notable example of the environmentalist bias in behavioral science is suggested by the concept of socialization (political or otherwise), typically understood as the "process" by which children (or adults) are led to conform to the standards of society (White, 1981b, 1992). In this formulation, children are viewed as malleable and passive and the process is unidirectional rather than genuinely interactional. These assumptions effectively deny "the animal" in us, that is, the genetic predispositions that we all begin life with and that may resist or even counteract environmentalist pressures. Strong-willed, left-handed children, for example, might successfully fight conforming to a right-handed regimen. Keeping in mind, however, the Popperian position that nothing in the sciences can be definitively proven (only disproven), an environmentalist position may be equally assumed in such a case without ever being demonstrated. In this book an interactional approach is proposed that is less counterfactual than either an environmentalist or hereditarian one.

In one crucial sense, environmentalist critics of biologically grounded approaches to the study of humans are more profoundly right than they may realize. Kitcher (1985), in discarding pop sociobiology, for example, holds out the possibility of a "serious human sociobiology" but notes that "in the case of *Homo sapiens*, the first rigorous analyses still awaits us" (p. 436). But what Kitcher's critique (among others) tells us is that a rigorous human science may be an impossibility. Identical twin studies constitute about as rigorous an analysis as human science is likely to attain, and if their findings on intelligence are suspect, then no findings are ever beyond suspicion (although it is noteworthy that findings that do not impinge upon egalitarian ideals are much more likely to be accepted).

For what Kitcher and others validly point out is the imposing complexity of genotype-environment interaction. Hirsch (1972) demonstrates, for example, that "ten genotypes in 10 environments generate 10^{114} possible kinds of interactions." He concludes, "Since the characterization of genotype environment interaction can only be *ad hoc* and the number of possible interactions is effectively unlimited, it is no wonder that the long search for general laws has been so unfruitful."

What such critics as Kitcher and Hirsch have wrought is an analytical sword so sharp and uncompromising that it can effectively cut through practically any generalization concerning any higher animal. It is also a double-edged sword: It can just as readily be applied to their own pronouncements, which when they concern human affairs usually take on an environmentalist tenor. Thus, on the one hand, Kitcher (1985) is at pains to show that "the idea of eyes and noses—or more pertinently, talents and dispositions—being handed down across the generations is a myth" (p. 21). There are, after all, simply too many problematic, environmentally influenced steps from gene to phenotype. Kitcher, on the other hand, goes on to state that "the aspirations and attitudes of individuals are typically shaped by the institutions of prior generations" (p. 30). Note that his observation has the same easy plausibility as the previous assertion that we may inherit our eyes, noses, or talents. But also note that Kitcher refrains from subjecting it to the same probing scrutiny. If he were to do so, he would find that each individual in a community, with his own unique genotype, may respond differently to the "same" social influences. There may not therefore by any "typical" response, a typological notion that would be immediately suspect if it were applied directly to a statement that stressed genetic influences. Moreover, one does not have to swallow whole Trivers's (1981) theory of parent-offspring conflict to concede the possibility that some children, especially during adolescence, may resist adult authority. In short, the same kind of cutting scrutiny that makes ribbons of loose general observations concerning the role of heredity can do the same when applied to the role of environment. (Hereditarians might provide a counter-demonstration in skill by completely "demolishing" the idea that environmental influences are a significant factor in human affairs.)

Furthermore, such an analysis is not in error because it correctly points to the complexity of the subject matter of human science and therefore to the likelihood that this focus of study will ever remain as much art as science. A rigorous study would probably conclude that nothing much influences anything and so presumably much of what happens is due to chance—a conclusion not far off from that of Jencks et al.'s *Inequality* (1972).

Environmentalist and hereditarian positions, moreover, deny the reality of politics as an autonomous activity. By reducing all conscious activity to epiphenomena dependent on, in the one case, environmental contingencies and, in the other, genetically programmed traits, all political interests and ideas are accordingly accounted for; they have no independent causal force of their own. The give and play of politics gets reduced to a set of internal or external factors; the political actor operates mechanistically as a product of either environment or genes. The mentalist perspective that is offered here restores to politics its autonomy.

As already mentioned, a typological approach tends to underplay both the "animal" and the "political" aspects in *zoon politicon*. In Mayr's classic formulation (1963), a typological view stresses the average "type" within any population, in contradistinction to a populational perspective, which stresses the wide range of individual variability that is likely to characterize a population. Within a populational framework, it is the diversity that is the reality and the average type at best a statistical artifact. But a populational emphasis on diversity gets in the way of the effort within behavioral science to formulate general laws that may be extrapolated into the future as scientific prediction. As the evolutionary process may defy the formulation of general laws and scientific prediction, so the political process, encapsulated within human evolution, may likewise elude the lures of a hard science. But the insistence of the behavioral scientist in typologizing the subject matter has impoverished our understanding both of the "political" and the "animal," that is, of the very stuff of politics. The political process is an outcome of competing forces, as represented in part by biologically unique individuals. To attempt to grasp it only by abstracting from it those factors that are common to it—or large portions of it—is to lose its essence. It would be as if one attempted to understand the Compromise of 1850 in American politics without alluding to Webster, Clay or Calhoun, or the Iraqi-Kuwait Crisis of 1990–91 without mention of George Bush or Saddam Hussein.

A typological view does violence to biological reality—and hence the "animal" within us—by ignoring or minimizing the biological individuality characteristic of each of us, an individuality noteworthy in the most "exceptional" individuals, including the "heroes in history" that Sidney Hook spoke of several generations ago. Can modern Russian history and politics be understood without reference to Lenin, Stalin, and Gorbachev and to the unique constellation of genetically influenced traits that they manifested?

In order to correct the glaring deficiencies that seem to underlay contemporary behavioral science, an IMPish alternative approach is proposed: a paradigm that is interactional, mentalist, and populational. Let us look at each of these characteristics in turn.

INTERACTIONAL

Consistent with Mayr's first "great truth of population zoology," heredity and environment "influence nearly every trait" (Mayr 1963, p. 650). The alternative is to believe that the environment by itself or the genes by themselves wholly determine each of our characteristics; surely either of these positions is a less defensible proposition. An interactional view seems to me to be the most plausible and appropriate (White 1972). No single interactional perspective, however, can be mechanically applied, for genetic and environmental influences will vary by physical or psychological trait, by human population, and through time (cf. Dobzhansky and Boesiger 1983; ch. 7). The relative impact of genetic and environmental factors on specific human traits and actions therefore will always remain an open empirical question.

This is where the "genes" of the title come in. Granted our genetic endowment does not completely program our actions. Nonetheless, it is possible, with all due respect to an environmentalist position, to see our genes as leading us from the start in certain directions: to be left-handed instead of right-handed; to be phlegmatic instead of temperamental; to be quite concrete instead of abstract in one's cognitive orientation. In the chapters to follow, the political implications of possibilities such as these will be discussed.

In this discussion the concept of self-selection will loom large: the idea that our genetically based propensities and capabilities will lead an individual during the course of his or her life to seek environments which he or she finds congenial. One person's pleasure is another person's pain; what one person seeks, another avoids. But out of a welter of environments and heredities, one constant emerges: the search of each individual for a favorable surrounding.

Although present throughout all of animal life, this process of self-selection plays its most dynamic and distinctive part in human society. Consequently, it supports what C. D. Darlington has referred to as the illusion that individuals and populations "are moulded by an environment which they have actually chosen," for "the environment with which a man's genotype reacts is itself, in its most profound character, the creation of men of like genotype." There is considerable merit, therefore, in Darlington's characterization (1969, pp. 252, 269, 270) of the environmental determinism running throughout much of contemporary social science as a "causal inversion" and "systematic fallacy"—provided one avoids the opposing snare of a genetic determinism.

One needs to examine the role of self-selection on both the population and individual level. If similarities in environmental setting characterize populations of individuals, especially over time, it is probably because such individuals start with similar genotypes. They consequently proceed to seek out similar environments, including individuals like themselves, through bonding and assortative mating for example. If their attempts are successful, they will share a common environment and similar phenotypes (see chapter 2). And if the environment is, in fact, favorable, then such effective self-selection will enhance the evolutionary chances for survival of the population.

Indeed, some capacity for self-selection is vital for the survival of any species. As Dobzhansky (1970, p. 25) points out, "No species is specialized to live in just a single environment. Since any environment constantly changes, such a species would not endure for long.'' It follows, therefore, as Mayr (1963, pp. 246–47) advances, that ''an ability to choose the 'right' background would tremendously increase the probability of survival'' for a genotype. Mayr further reports that ''evidence of such habitat selection by genotypes is beginning to accumulate.''

Aside from the problem of its explication, the widespread selection of habitats by animals is well known. In *Biology and the Future of Man*, (1970, pp. 380–81; emphasis in the original) Handler draws the following picture.

> The observer of animals in nature is aware that each species of animals is to be found in a limited habitat. . . . What keeps these animals from going into the ''wrong'' habitat? The problem is most clearly shown by those animals, such as many birds and fishes, which migrate every year for great distances from their birthplaces, and which, when they return home at the beginning of their breeding season, unerringly *choose* to settle in the characteristic habitat of their species, even though other habitats seem equally available.

Insofar as purposive behavior and lower forms of consciousness characterize animal life in general, the evolutionary process no longer can be viewed as blindly mechanistic. According to Dobzhansky (1970, p. 93), ''Even lower animals move where conditions please them.'' Of course not all attempts at self-selection will succeed, and indeed Mayr (1982, p. 358) points out that every ''major shift of habitat is an evolutionary experiment'' and that ''most of them are failures.'' Nevertheless, Mayr (1982, p. 573) asserts that ''What is most remarkable is that bearers of different gene arrangements not only have a different fitness in different niches but also the behavioral capacity to search out the right niche.''

To the extent that organisms can take actions that are not wholly constrained by genetic or environmental factors, they may affect their own chances of survival. The organism then plays an active role in the evolutionary process and ceases to be wholly the passive product of blind external forces commonly associated with the term ''natural selection.'' Rather what has been termed ''organic selection'' emerges, a theory whose roots may be traced back to C. H. Morgan (1896), J. M. Baldwin (1897) and Hardy (1965). This theory asserts, as Popper and Eccles (Popper and Eccles, 1977, p. 12) put it, that

> all organisms, but especially the higher organisms, have a more or less varied repertoire of behavior at their disposal. By adopting a new form of behavior the individual organism may change its environment. . . . In this way, individual preferences and skills may lead to the selection, and perhaps even to the construction of a new ecological niche by the organism . . . thus the activity, the preferences, the skill, and idiosyncracies of the individual animal may indirectly influence the selection pressure to which it is exposed, and with it, the outcome of natural selection.

In these and other ways humans also influences their own evolution. Boehm (1978, p. 80) argues that efforts to control the environment have made and will continue to make a difference in the adaptive possibilities, probabilities and outcomes experienced by human groups. These outcomes are of enormous potential importance to the continued adaptive effectiveness of human populations—and possibly to their very survival.

On an individual level, Scarr and McCartney (1983, p. 424) propose a theory of human development "that emphasizes the role of the genotype in determining not only which environments are experienced by individuals but also which environments individuals seek for themselves." They note that genotype-environment interaction may take three forms: passive, evocative, and active.

Passive. This is especially relevant for the very young, who inherit both their genes and their family environment from their parents. Since the parents' own genes influence the environment that they provide, it is clearly extremely difficult to disentangle the effects of genetics from the environment, so closely are they intertwined. But whatever the case, the impact of both influences upon the child is bound to be profound and constitute an inheritance that is a "given," that is, the child has no say over the initial contribution of his or her parents; he or she is the passive recipient.

Evocative. The child, with his or her unique, genetically influenced traits, will evoke differential responses from his or her family and from the environment in general. But it should be added here that even if parents were to treat their children uniformly alike (or to try to), their biological individuality would still assert itself. As Bouchard et al. (1990, p. 227) note, "Different individuals pay different attention to or respond differently to the same objective experience."

Active. "People seek out environments they find compatible and stimulating. We all select from the surrounding environment some aspects to which to respond, learn about, or ignore. Our selections are correlated with motivational, personality, and intellectual aspects of our genotypes" (Scarr and McCartney 1983, p. 427). This third form, they proceed to argue, "is the most powerful connection between people and their environments and the most direct expression of the genotype in experience." This active form of genotype-environment interaction, or self-selection, will inform much of the theoretical content of this work.

One critical caveat needs to be repeated: A genetic propensity or potential does not assure its fulfillment. The environment always plays a vital, ever-present role. Thus slavery made it nearly impossible for any slave to attain his or her potential. It also remains to point out, though, that slavery did not eliminate this biological capability; it rather prevented its realization. In general, then, while the environment remains a potent, ever-present influence on our lives, it never negates our biological natures, even if it may inhibit their realization. The chapters that follow will elaborate this position. In the discussion of a populational perspective in this "Introduction," I will allude to it in connection to individual responses to the horrors of Auschwitz.

It remains to note that an interactional perspective has indeed gained greater acceptance within the fields of psychology and the social sciences in recent decades (cf. Degler 1991). Political science has remained relatively impervious to this trend, but even in this discipline a growing biopolitical movement, culminating in the formation of the Association of Politics and the Life Sciences, can be discerned (see, e.g., Somit 1976; Wiegele 1979; Corning 1983; Watts 1983; Masters 1989; and Schubert 1989).

Although those of us in "biopolitics" basically share an interactional approach, there are perhaps several differences in emphasis that distinguish this particular work from those of my colleagues. The first is the stress on self-selection as a central concept. The second and third differences involve the explicit adoption of a mentalist and of a populational framework.

MENTALIST

According to R. W. Sperry (1983), humans are endowed with a self awareness that is an emergent property of the brain and that enjoys causal potency. One should add that Sperry's views on consciousness and the brain are monist rather than dualist (Popper and Eccles [1977] are both mentalist and dualist, if not trialist), and scientific rather than religious or vitalistic. A mentalist view, which we adopt here, indeed enjoys increasing support from leading neuroscientists with otherwise disparate opinions (Eccles 1977; Granit 1977; Pribram 1981).

A mentalist, in contrast to a mechanistic, outlook grants a causal role to the purposive or teleonomic activity of the individual; our conscious aims are here more than mere epiphenomena. A mentalist perspective does not deny the influence of irrational or of blindly internal or external factors of genetic or environmental origin. It simply denies that all conscious thoughts can be understood solely by reference to lower-level causes, whether genetic, physiological, or environmental in nature.

The "brain" of the title—the preeminent human organ—plays a vital role here. It is responsible, in the first instance, for our human awareness, but it also provides the neurophysiological context within which that awareness will operate. Our neurophysiology, for example, will include the level of serotonin and MAO (monoamine oxidase) that will, in turn, influence our moods, such as whether we will feel depressed or impulsive. These neurophysiological factors are of course directly influenced by our genes, and the emerging field of developmental neurobiology will likely come to trace these connections with greater precision.

Of course, environmental influences are always present and always play a role. But in and of themselves they do not determine our responses, which remain conditioned by our biological inheritance. The diathetic stress model of mental disorders such as schizophrenia and manic-depression provides an illustration. Under optimal environmental circumstances such mental disorders may be minimized or even avoided. But under traumatic circumstances (constant child abuse, for example), the disorder may be triggered and developed in severe form. The symptoms, however, will vary according to individual predisposition, whether schizophrenic, manic-depressive, etc. Indeed, there will be individual variation within each general disorder, as indicated by a populational approach, to be discussed shortly.

Some exceptional individuals, moreover, may endure a traumatic childhood without any notable deficiencies in their mental health. They may exhibit the "invulnerable" personality described in Neubauer and Neubauer (1990, Ch. 7), that is, the individual whose innate tendency to respond in an upbeat, resilient manner to the vicissitudes of life actually begins at birth (cf. the "autotelic personality" in Csikszentmihalyi 1990, pp. 92–93).

Clearly such personality differences are laden with political consequences, and the ensuing chapters will attempt to elucidate these. Just a brief example will suffice at this point. In their classic cross-cultural political survey, Almond and Verba (1965) noted that adults who remembered having had no opportunity as children to participate in the decisions taken in their homes and schools actually, on average, attained a higher level of subjective citizen competence than those adults who remembered having had such an opportunity but who failed to avail themselves of it. (A third group, who had both the opportunity to participate and took advantage of it did indeed attain the highest level of subjective citizen competence.) Since this subjective sense actually translates into overt political activity in many cases, it is a significant finding, but also a surprising one from a wholly environmentalist perspective. "Whence this strong ego?" one might ask, in reference to the adults who developed a strong sense of political efficacy, the feeling that their voices can count in political affairs, despite having grown up in environments that discouraged their active participation in key decisions that may have affected them.

> Whether or not a child feels free to protest or actually does protest in family or school is not determined solely by the adult-created authority structure of the family or school. Rather, family or school participation and political participation may both be affected by the extent to which the child has a strong ego. (Almond and Verba 1965, p. 292)

Two points need to be made here. First, the "strong ego" referred to may include a cognitive dimension that assumes the mentalist position taken here. Parts II and III will elaborate this possibility. Second, important individual differences may exist with respect to this cognitive dimension and the following section touches on these.

POPULATIONAL

Since I continue to be influenced, as I had in my original interactionist formulation, by Ernst Mayr's "two great truths of population zoology" (1963, p. 650), I reiterate them here: "No two individuals are alike" and "Both environment and genetic endowment make a contribution to nearly every trait." The second "truth" summarizes an interactional viewpoint and the first a populational one. If all individuals are biologically and, excepting monozygotic twins and supertwins, genetically unique, and if indeed the environmental circumstances that attend each individual are also likely in some critical aspects to be unique (if for no other reason than the uniqueness of the individuals surrounding one and thereby forming a key part of one's environment), it becomes clearly a mistake to ignore or minimize individual variations, as behavioral scientists and those wedded to the goal of a hard systematic human science are in fact prone to do. Rather than typologizing or over-generalizing, one will remain truer in his or her perception of the human condition if he or she fully acknowledges the wide range of individual variation that will likely characterize any human population. Whether we speak of physical stature, intelligence, or shyness, we speak of polymorphic traits that will vary widely in any large group. (Dimorphic traits, such as certain blood types, vary of course in a more circumscribed and discrete fashion.)

A populational approach can be justified, in fact, on grounds of both science and policy. The desire to attain a science of general laws may be inappropriate to an understanding of human affairs; it may in fact impede our understanding. The urge to typologize (which itself may have an innate neurobiological basis) may produce what Mayr (1972) has termed "retarding concepts." A prime example of this in evolutionary biology involves the idea of evolution itself. According to Mayr (1972), the formulation of the evolutionary process was hindered by the then prevalent Newtonian mentality in science, which aimed for universality, repetition, and permanence—attributes that are perhaps not even applicable to the physical universe over its entire lifespan, but which certainly appear inapplicable to animal life. An understanding of the evolutionary process involves the recognition of genetic variability and irreversible change (cf. Gould 1990). Insofar as human life is part of such a process, it too must be described in these terms. To abstract from human life only that which is common and unchanging is to leave out its essence and, consequently, its politics.

Human differences will emerge under the most uniform of conditions, even if those conditions happen to be extremely oppressive. Frankl (1984, p. 178; emphasis in the original); provides the following observation of his experience at Auschwitz:

> Sigmund Freud once asserted, "Let one attempt to expose a number of the most diverse people uniformly to hunger. With the increase of the imperative urge of hunger all individual differences will blur, and in their stead will appear the uniform expression of the one unskilled urge." Thank heaven, Sigmund Freud was spared knowing the concentration camps from the inside. . . . *There*, the "individual differences" did *not* "blur" but, on the contrary, people became more different; people unmasked themselves, both the swine and the saints.

Admittedly, Frankl has in mind here differences in moral conduct in particular (though, as will be argued in chapter 3, these too may have a biological underlay), and most certainly he did not equate exemplary moral conduct with survival:

> There was a sort of self-selecting process going on the whole time among all of the prisoners. On the average, only those prisoners could keep alive who . . . were prepared to use every means, honest or otherwise, even brutal force, theft, and betrayal of their friends, in order

to save themselves. We who have come back, by the aid of many lucky chances or miracles . . . know: the best of us did not return. (Frankl 1984, pp. 23–24)

The larger point, however, is that the social environment—as uniformly oppressive as it was—did not produce a uniform response: "Even though conditions such as . . . insufficient food and various mental stresses may suggest that the inmates were bound to react in certain ways . . . it becomes clear that the sort of person the prisoner became was the result of an inner decision, and not the result of camp influence alone" (Frankl 1984, p. 87).

Here Frankl is clearly stressing subjective factors in a way wholly compatible with the mentalist position described earlier. It is also possible to infer from Frankl's observations an underlying biological dimension. Aside from any amorality on his part, it is not hard to see two genetically influenced traits as assisting Frankl's survival. The first is an affective trait; Frankl (1984, p. 28) refers to his "inborn optimism" which "has often controlled my feelings even in the most desperate situations." The second is cognitive, in his case an exceptional ability to think in abstract as well as concrete terms and hence to distance himself—both in time and space—from his immediate hell:

I became disgusted with the state of affairs which compelled me . . . to think of only such trivial things. I forced my thoughts to turn to another subject. Suddenly I saw . . . [i]n front of me . . . an attentive audience on comfortable upholstered seats. I was giving a lecture on the psychology of the concentration camp! All that oppressed me at that moment became objective. . . . By this method I succeeded somehow in rising above the situation . . . and I observed them as if they were already in the past. (Frankl 1984, pp. 44–45)

In short, it is plausible that genetically influenced capabilities and predispositions played a key part in the variation of individual response that Frankl observed in Auschwitz.

Unhappily, efforts to fashion a uniform environment along oppressive lines are already a well-documented part of human history. Fortunately, these initially all-too-effective efforts can be sustained for only so long. As Sinnott (1945) argued in his classic "The Biological Basis of Democracy," an inevitable human diversity defies the strongest attempt of any totalitarian regime to impose a monolithic party line upon a large population over a number of generations. His analysis serves the events of 1989 in Eastern Europe well.

It is clear that the whole provocative area of the implications of a populational perspective for public policy has now been broached. The introduction of genetic factors into human science has been understandably questioned in view of the racism justified by social Darwinism and Hitlerian fascism. Yet a look at the most extreme results of human prejudice—slavery and genocide—will indicate that they in no way depended upon the introduction of genetic factors, especially as interpreted within a population framework. On the contrary, both slavery and genocide can be shown to rest on justifications that are, above all, typological in their formulation and may not even include a critical biological component. As Sowell (1983, p. 144) notes, for example, the institution of human slavery long predated the Darwinian theory of evolution and many of its presumed social implications during the later half of the nineteenth century; if anything slavery was well on its way toward abolition by that time and nothing in the new evolutionary theories seems effectively to have forestalled its demise.

Of the two most notorious incidents of genocide in modern times—under Stalin and Hitler—that under Stalin, the "liquidation" of millions of so-called kulaks under a policy of violently enforced collectivization, had no biological element in its justification. Rather, in a blatantly typological formulation, all those peasants who actively opposed the Communist-imposed policy of collectivation were pejoratively labelled "kulaks" and were therefore to be treated collectively as "enemies of the people."

Indeed, Conquest (1990, p. 27) declares that many Party members under Stalin who would react with "horrified resistance" to "the execution of prominent Party members, to 'shedding' the blood of Bolsheviks," earlier had "cheerfully massacred the whites, and at least uncomplainingly starved and slaughtered the peasantry." Conquest sees this split morality as the "*dvoeverye*—double belief—of the Party 'moderates' "—"a double standard of morality comparable to the attitude of sensitive and intelligent men in the ancient world to slaves or the French nobility of the eighteenth century to the lower classes." Such a double morality is notable in the formulation of its typological dichotomy.

Hitler's racism and subsequent genocide did have a key biological component at its center. But it was its explicit typological formulation that provided the chief motive force. What Hitler most wanted to achieve was a biologically "pure" Aryan race. Hitler did not, for example, believe that the Jews were "genetically inferior"; their only crime was constituting an alien element and therefore impediment to Aryan "racial purity." Thus Hitler (1954, p. 313; emphasis in the original) stresses as the purpose of the state "*the preservation and advancement of a community of physically and psychically homogeneous creatures*." It is this typological "racial purity" that must be protected: "With satanic joy in his face, the black-haired Jewish youth lurks in wait for the unsuspecting girl whom he defiles with his blood, thus stealing her from her people. . . . It was and is Jews who bring the Negroes into the Rhineland, always with the same secret thought and clear aim of ruining the hated white race by the necessary resulting bastardization." (Hitler 1954, p. 325).

Almost all nationalistic excesses can be viewed as resulting from the attempt of one ethnic group typologically defining and attempting to impose an alien or inferior status upon another. The Turkish-Armenian, Sino-Japanese, and Indian-Pakistani ethnic conflicts illustrate this unfortunate fact.

Ironically, the attempt to disregard genetic factors in human science typically takes shape within a typological formulation that also does violence to our understanding of human affairs and to our conduct of public policy in much more subtle ways. Based upon friendly, but argumentative, meetings with my brother-in-law in Berkeley, California, I have "typologized" his and many other left-of-center views as constituting "the Berkeley mentality." Basically, it consists of the opposite of most of the positions taken here and can be characterized as environmentalist, egalitarian, typological, and utopian. A brief explanation of these descriptive terms and why they may be more damaging than constructive in matters of public policy follows.

The aim of achieving an egalitarian utopia provides the key to the problem. Politically, this goal translates as a civil society in which each member not only has an equal opportunity to participate but in fact does equally participate; in other words a participatory democracy in which the political influence of each citizen is as directly transmitted as possible and is roughly equivalent.

This vision operates upon the neo-Rousseauian assumption of the perfectibility of man, an assumption that permeates contemporary third-school psychology. It tends to see human nature as largely undefined and malleable and therefore each individual as potentially the equal of another. Thus it emphasizes the environment, usually expressed historically through a particular socioeconomic system, as the primary source of an individual's standing in society. Change "the system" and you change the individual.

Granted the critical role of the environment, an overestimate of its causal efficacy has a distorting effect. This distortion ensues precisely as the result of its typological formulation. If the particular socioeconomic system is judged not to promote an egalitarian utopia, then most of its members, especially those seen as most "disadvantaged," are viewed indiscriminately as "victims." As was once stated in response to the description of the Baltimore antipoverty program in the 1960s by Bachrach and Baratz (1963), such a radical critique of society comes seriously close to a "social Darwinism in reverse: "Poor blacks can never be less than bright or anything but victims; the white power structure, on the other hand, may on occasion be inept and is in any case oppressive" (White 1972, p. 1240). The typological categorization of an entire group or class of individuals as "victims of the system" ultimately robs them of their individuality as well as divests them of any personal responsibility.

Conversely, the typological vision of an egalitarian world in which the political and economic status of each member is equivalent has similar consequences. (This is so even for personal responsibility; thus take the admittedly more extreme example of a Marxist-communist society: If in theory everyone owns the means of production, then no one in particular bears any personal responsibility for its use, or misuse, as when uncontrolled environmental pollution results.)

In short, this egalitarian utopian vision is also environmentalist and typological in its underpinnings. It ignores Dobzhansky's basic insight (1973) that insofar as an equality of opportunity is possible and insofar as an environment can be made uniform, the role of heredity and, therefore, the reality of individual differences becomes, if anything, even more central. Therefore, inequalities would persist precisely because of the persistence of genetically based individual variation.

As Sowell (1983, p. 244) notes: "Human differences are often assumed away in social theories either because they would needlessly complicate the exposition or because they are presumed to be negligible or subject to change by 'society,' which takes each new individual as a blank page on which it can write its social message."

If human differences, and hence inequalities, persist, must that fact constitute a problem? In Sojourner Truth's famous speech "And Aren't I a Woman?" (cited in Bernard 1967, p. 178), she asks: "What's intellect got to do with women's rights or black folks' rights? If my cup won't hold but a pint, and yours holds a quart, wouldn't you be mean not to let me have my little half-measure full?" In other words, if one applies a populational perspective to public policy, it is irrelevant what the sex or race of the individual is; only the individual and his or her capabilities matter. This would hold even if the larger groups of which that individual is a part might differ, on the average, in particular capabilities. Indeed, Sowell (1983, p. 244) insists that, whether for cultural or genetic reasons, or a combination of the two, "the most obvious fact about the history of racial and ethnic groups is how different they have become and still are." Here in particular the need to rely on a populational approach becomes clear, that is, to acknowledge explicitly and constantly the wide range of human variation that will inevitably characterize any large population and to avoid typologizing all of its members in line with some statistical average.

But the liberal egalitarian may typologize in his or her own way with subtle but negative consequences. The insistence on viewing all individuals, and hence all groups, as potentially equal introduces its own iatrogenic social problems. That is to say, the insistent need to compare measures of group achievement as if any inequality were a problem thereby creates a problem. Why must there be as many champion female chess players as male or as many white champion basketball players as black—provided only that no individual is denied equal access to the top positions? The unremitting emphasis on group differences and their persistence unwittingly, perhaps, graphically typologizes groups and their individual members.

This discussion of the public policy implications of a populational approach is by its nature political because it treats alternative ways of organizing a social policy in order to deal with human inequality. In so doing, it emphasizes that a populational approach implies the inevitability of politics. As in James Madison's analysis of the origins of "faction" in *Federalist No. 10* (issuing first and foremost from the "diversity in the faculties of men"), a populational emphasis views an underlying human diversity as giving rise to a multiplicity of aims and interests that, at the level of societal decision, will in some cases produce clashes or compromises and, either way, politics. The seeds of both "faction" and politics are "thus sown in the nature of man."

But just as Madison also thought it possible "to refine and enlarge" the views of most citizens by "a chosen body of citizens distinguished by their wisdom, patriotism and love of justice," so a mentalist position would allow for the possibility that political ideas and public policies might attain a certain autonomy and be evaluated on their own merits.

FRAMEWORK

The remainder of this volume will incorporate an IMPish approach to a number of areas of social and political importance. Given the relative youth of the life sciences directly concerned with human action—for example, neurobiology and behavioral genetics—much of what follows is speculative. Many statements may be taken as hypotheses requiring further exploration and testing rather than as definitive conclusions, which in the social sciences are unlikely in any event.

Moreover, there is no pretense here to be comprehensive. The whole of political life could presumably be treated within the context of an IMPish perspective but assuredly not within the confines of a singe volume. While what follows is indicative of some of my own research interests, it may prove suggestive in its approach and concepts in ways that may apply to other areas.

The overall organizing concept of this book is that of self-selection. If in a social world marked by endless variability and unceasing flux, there is one constant, it is the persistent effort of each of us to find a suitable niche within the larger scheme of things. This effort is elaborated in the three parts that constitute this work. Part I treats the apparently universal self-selecting propensity of each of us to find congenial companions and the political implications that follow. The examples used in chapters 2 and 3 to illustrate this analysis involve the recent American scandals of Watergate and Iran-Contra. Most of the examples used come from American experience, but references are also made to experiences elsewhere, both geographically and historically, to augment and universalize the discussion.

The emphasis in Part I on the formation of congenial bonds and networks, for example, appears to apply to the Russian political experience. Ra'anan and Lukes (1990, p. 11) note that "the importance of personal contacts can hardly be overestimated in any political system, open or closed." They also add, "But the top Soviet elite seems to rely on connections between itself and new candidates more than most political systems in modern history." Referring to the then top Soviet management elite, Klugman (1990, p. 90) writes that "executives everywhere want to work with known entities; and in the Soviet Union where informal relations are so important—this desire is especially strong."

According to Yeo-chi King (1991, p. 63), "No one who has lived in mainland China, Taiwan, Hong Kong, or any other overseas Chinese society could be totally unaware of a social phenomenon called *kuan-hsi* (personal relationship). It is striking to note that although the Chinese Communists in mainland China have repeatedly launched rigorous campaigns—including the Great Proletariat Cultural Revolution—to attempt to uproot 'feudal' elements of the Chinese culture," Yeo-chi King continues, "kuan-hsi in particular" remains "strong." Indeed, Yeo-chi King (1991, p. 64) cites Fox Butterfield as asserting that the personal connections of the Chinese "operate like a series of invisible threads, tying Chinese to each other with far greater tensile strength than mere friendship in the West could do."

Let us also take note of the Japanese experiences with its emphasis on *kone* or "connections" and *jinmyaku* or larger networks in its political and commercial life. Indeed, Wolferen (1989, p. 110) views *jinmyaku* as being "much more widespread, and of incomparably greater importance, than old-boy networks in the West." Nakane (1970, pp. 120–721) also stresses the formation of informal group ties in Japanese society, which she sees based upon "congenial friendship, long and stable association through work; common backgrounds," among other major factors. She goes on to note that "among fellow members a single word would suffice for a whole sentence." Such intimate rapport is quite similar to that enjoyed by identical twins. It remains only to make explicit the possible biosocial underpinnings for the rapport, as well as to draw out the political ramifications.

Part II treats self-selection within the context of local and cosmopolitan orientations. Chapter 4 elaborates a local-cosmopolitan continuum, and the following chapter examines the role of local and cosmopolitans in the fight over the ratification of the U.S. Constitution.

Again, the applications here may extend beyond the borders of the United States. Nakane (1972, p. 23), for example, discusses the "strong localism" in Japanese society, for which she employs the Japanese terms *inakappe* or "unsophisticated countryman." She adds that the English terms *local* "fits the situation admirably."

Part III extends the cognitive dimension implicit within a local-cosmopolitan continuum to make it more or less co-extensive with a concrete-abstract continuum in intellectual ability. Here we enter into an area that is often avoided precisely for its thorny ideological overtones. Indeed, from a hereditarian and especially from a typological perspective, the implications may be quite troublesome. A populational-interactional view, on the one hand, avoids any racist and sexist overtones, but on the other hand, an emphasis on self-selection must also emphasize the centrality of cognitive ability or, in a general sense, intelligence. Indeed, Sternberg (1988, pp. 65, 72) defines *intelligence* as the "purposive adaptation to and selection and shaping of real-world environments relevant to one's life." And although Sternberg (1988, p. 75; emphasis in the original) stresses the more egalitarian and environmentalist basis for intelligence, he does acknowledge the "overwhelming evidence that suggests that intelligence is *in part* genetically determined."

Based upon a cognitive framework, chapter 6 examines individual-environment interactions with a view toward the differing forms of alienation that they may produce. The final chapter (chapter 7) combines this analysis with that on local-cosmopolitan self-selection and applies it to the role of women in contemporary American society. The defeat of the ERA in the United States serves as the focal point of the analysis.

The biopolitical analysis found here is bound to raise serious questions in research, theory, and policy. The easiest and most likely response for someone of a different mindset is to view this entire enterprise as speculative, unsubstantiated, and, in any event, premature and, therefore, to be safely ignored. But those more mindful that the foundations of the behavioral approaches still in vogue are at least equally as shaky will view this effort with a more receptive eye. They will be willing to consider statements offered here as tentative hypotheses to be explored and tested rather than to be summarily rejected. They may also see this effort as premature, but, perhaps, in the sense of serving as an introduction to new, more promising avenues of social and political investigation.

Part I

Brains, Bonds, and Politics

Chapter 2

Brains, Bonds, and Bureaucracy: Public Administration from a Neuropolitical Perspective

> ''It appears, then, Menexenus and Lysis, that passionate love, friendship, and desire happen to be for what is akin, at it seems.''
>
> They [both] assented.
>
> ''You, therefore, if you are friends to each other, are by nature in some way akin to each other.''
>
> —Plato, *Lysis* 22e

> For without friends no one would choose to live. . . . Parent seems by nature to feel it for offspring and offspring for parent, not only among men but among birds and among most animals it is felt mutually by members of the same race, and especially by men, whence we praise lovers of their fellow men. We may see even in our travels how near and dear every man is to every other. Friendship seems to hold states together, and lawgivers to care more for it than for justice.
>
> —Aristotle, *Nicomachean Ethics*, 1155a 5–25

> Where we seek to devise mechanisms for inducing people to work effectively together, we are not fighting chaos or fending off entropy, we are providing an outlet for hereditary drives . . . there is a genetic social force in human organization that we have not considered adequately.
>
> —Luther Gulick, ''Democracy and Administration Face the Future''

How do organizations select their members and with what results? This question is addressed within a neuropolitical context. In essence it is asserted that individuals who, by virtue of similar brains, share similar interests and outlooks tend to bond and to form networks and hence organizational clusters. Basically, it is argued that individuals form bonds on genetic, environmental, and ultimately neurobiological grounds. This bonding process is examined within an organizational context.

Next, the implications of this analysis for the study of public administration is explored. Three major conclusions are drawn:

1. Friendship ties that are inherent within social networks may independently influence the process of organizational personnel selection and therefore the character of the organization itself.
2. Inasmuch as the formation of organizational clusters reflect genetic, environmental, and neurobiological factors, what in the behavioral sciences is often viewed to be primarily environmentalist in etiology has to be conceived rather as multileveled and, following Corning (1983), synergistic in character.
3. In examining social network and organizational ties, there is ultimately no substitute for examining the unique individuals who occupy specific roles, especially at the highest organizational echelons.

Each of the critical terms used—*bonds, networks,* and *clusters*—will be defined as the chapter proceeds. Finally, although the focus of the analysis is on public administration, references to the private sector are made when the points under discussion apply also to the governmental realm.

THE INTERACTION-MENTALIST-POPULATION (IMP) PARADIGM APPLIED

> Father and son may be of entirely different dispositions, and brothers also. He is my son, he is my kinsman, but he is an unsociable man, a knave, or a fool. And then, the more there are friendships which law and natural obligations impose on us, the less of our choice and free will is in them. And our free will has no product more properly its own affection and friendship. . . . In the friendship I speak of, our souls mingle and blend with each other so completely that they efface the seam that joined them, and cannot find it again.
>
> —Montaigne, *Of Friendship*

> I lost my mother ten years before, and the eternal separation had been a bitter blow. But between my father and myself there existed a friendship altogether independent of natural affection, and perhaps even deeper. Our ideas were entirely alike on many questions, the mere approach to which electrifies the intelligence of certain men.
>
> —Berlioz, *Memoirs*

Brains, Bonds, and Self-Selection

To paraphrase C. D. Darlington (1969, p. 269), each individual seeks and selects the environment that fits him best. On an instinctive level such self-selection characterizes the animal kingdom generally (Mayr 1963; Dobzhansky 1970).

Humans, of course, share such deeply rooted, biologically based drives as hunger, thirst, and sex, but we qualify and may even transcend them (committing suicide in the process if need be) in the pursuit of other aims whose conscious status may make them preemptive rather than derivative. In any case, human beings seek, both consciously and otherwise, congenial environments; and since a key part of any environment is that which is composed of other individuals, we seek congenial companions (cf. Scarr and McCartney 1983).

The congeniality of our environments and therefore of our companions is ultimately defined by each individual for her- or himself. The old adage, "One man's meat is another man's poison," has some application here. Nonetheless, while the choices involved undoubtedly have a significant, and perhaps overriding subjective component, they may also rest upon objective grounds. To the extent that individuals share common or similar genetic and environmental backgrounds, the propensity to bond should increase.

The ultimate test of this proposition involves identical twins. "Twins, everywhere," writes Robert Brain (1976, p. 124), "are perfect friends":

> In African cosmologies and classical myths twins symbolize the fusion of the couple, the perfect dyad, the intimate union, Montaigne's one soul in two bodies. Mythical twinness, an almost universal element in cosmologies, may represent . . . the victory of order over anarchy.

Brain goes on to note that "The intimacy of mythical twins becomes the ideal friendship." He concludes that this idea is "a common theme in most cultures" and proceeds to illustrate this statement by reference to cultures throughout the world (see Brain, ch. 6).

Monozygotic twins are in fact found to be unusually close in their social relationships (see Koch 1966; Segal 1988).[1] The close social ties enjoyed by identical twins are clearly related to the genetic and environmental background that they hold in common. But this can perhaps be more directly traced to the fact that they share, for all intents and purposes, the same biological brain (although this is, of course, not so literally). As Segal (1988, p. 175; emphasis in the original) writes, "It is not the *fact* of shared genes that is critical for social bondings, but rather the *ways in which shared genes influence the brain to respond to objects and events in the environment*." With similar brains, they are likely to view the world alike, to share common tastes, and to enjoy similar pursuits. So it is that identical twins typically develop their own private means of communication in which a co-twin frequently completes the other's thought before it is fully expressed.

Neurophysiological tests indicate that identical twins share common brain wave patterns (see Lykken 1981). It may be literally true that when such twins enjoy each other's companionship they are indeed on the same wave length. More generally, in Karl Pribram's holographic model of the brain (1979, pp. 74, 80) human communication is literally, not just figuratively, predicated on the same brain waves: "The brain can instantly resonate to and thus 'recognize' wave forms. . . . We apparently need to get on the same wave length—literally—before we can understand each other." (From a mentalist perspective, the brain waves as such reflect, rather than determine, the ties of mutuality and friendship.)

Intuitively, we understand that members of the same species, with the same basic genetic heritage and therefore basically similar brains, will more readily communicate among themselves as opposed to among members of other species. Inasmuch as each brain—and certainly each human brain (Sperry 1983)—is unique, would it not follow that communication and hence friendship ties would be facilitated among those individuals most resembling one another in key neurobiologically-based intellectual and temperamental traits?

It is within this general context of neurobiological similarity, furthermore, that the adage "opposites attract" may apply. Thus Arthur Schlesinger, Jr. (1960, p. 226), in describing the New Deal friendship between Thomas G. Corcoran and Benjamin V. Cohen, writes that they "made an ideal combination." . . . "Corcoran's brashness supplemented Cohen's shyness, as his perpetual high spirits offset Cohen's occasional moodiness." It is true that individuals with certain similar traits, such as brashness, may tend to clash, so that in these cases opposite traits tend to complement one another (cf. Hinde 1979, ch. 7). But all that remains similar should not be overlooked. Both Cohen and Corcoran, for example, were brilliant, no doubt in the upper one percentile of the population. If they had, instead, been opposites in this respect—or even noticeably different—their attraction to each other would have been correspondingly diminished.[2]

Bonds, by which I simply mean social attachments, may take different forms and may be differently based.[3] Kin bonds are in effect inherited, as with parent-offspring bonds in general, and mother-child bonds in particular. From a sociobiological perspective, the mere quantity of the genes held in common suffices to link kin. Thus simply on sociobiological grounds alone, identical twins ought to be socially closer and indeed more altruistic toward one another than is the case with fraternal twins. This appears to be the case.

Nonetheless, it is conceivable that the quality as well as the quantity of the genes shared may matter when it comes to social bonding. Genes that influence physical traits may not be as critical as those that influence psychological ones (no doubt by influencing the design of the brain). We are all familiar with the fact that some family members are closer than others and that, in at least the case of parent-offspring relations, these closer ties occur independent of the numerical proportion of genes shared.

Brain (1976, pp. 16–17; emphasis added) observes that in some cases kin may in fact be defined and treated as such on the basis of freely chosen friendship ties.

In open societies such as our own, where kinship is organized multilaterally (not just down the male or female line), the number of possible kin is immense, and choice, outside the elementary family, determines the persons one treats as kin. *There would even be a case for maintaining that all kin relations within our kinship group are based on friendship and personal choice*. One chooses this or that uncle, this or that cousin, even this or that brother and sister to be friendly with. Unlike most other societies even kinship is achieved, rather than ascribed: *Kin do not interact because they are kin but through liking and common interests. Therefore kinship, for us, is like friendship in so far as it is a personal and private contract between individuals.*

What is critical is not so much the genetic or environmental dimension but rather the neurobiological one. As was just argued with respect to identical twins, it may be the sharing of a common outlook as the result of having relatively similar brains that is of critical significance.

Even parent-offspring relations may illustrate this point. Mary Soames (1979, p. 30) describes how her grandmother, Blanche Hozier, was "paralyzed with grief by the death of her favorite child": "Kitty and her mother had been kindred spirits, but Clementine's nature and her own were too dissimilar." George Jackson (1970, p. 49), in a letter from prison to his mother, wrote that his father did not share the same intellectual vision that they did, that he "just doesn't have the mental equipment." More than her other siblings, Shirley Chisholm (1971) shared the same high drive and intelligence of her father, welding so exclusive a bond between the two of them that the father's will excluded the others.

Such qualitative similarities in outlook, presumably neurobiologically based, are likely to underlie the formation of any freely chosen bond. Mating bonds, for example, are assortive rather than random and couples are likely to share similar characteristics, both psychologically and physically (Cattell 1982; Cattell prefers "assortive" to "assortative"). Of course if mating bonds are not randomly contracted, neither are they perfectly constructed. The high incidence of divorce and separation in most modern societies attests to this, and the fact of human genetic diversity alone insures a variety of traits that may not always be compatible in individuals who are otherwise attracted to one another. It is relevant to note that identical twins who marry other sets of identical twins and live and rear their children together enjoy unusually strong family relationships, indicated, in part, by a very low incidence of divorce and separation (Ravich 1972).

Ethnicity, Social Class, and Bonding

Members from the same ethnic background should, simply on sociobiological grounds, enjoy a relative advantage in bonding. Once again, however, it is necessary to transcend a merely mechanical genetic accounting. Creative individuals from the most widely differing backgrounds form international communities in the arts and sciences that transcend national, racial, and ethnic boundaries. Such individuals may bond and find themselves "on the same brain wave" quite independently of the precise proportion of genes they share. Nonetheless, such bonding is neither a random nor a wholly environmentally based phenomenon. Rather, individuals sharing similar genes for key psychological traits that are neurobiologically based and who also, like all individuals, share a common desire to select congenial environments, will form social attachments if the opportunity presents itself.

In a completely open society (if such was attainable), the genetic basis of class differences would be maximized. As Dobzhansky (1973, p. 29) writes, "It may chagrin some people to learn that increasing equality of opportunity enhances, not reduces, genetic differences between socioeconomic classes." With the attainment of a nearly perfect equality of opportunity, genetic heterogeneity within classes would be reduced, according to Dobzhansky (1973), and replaced by what he refers to as "aptitude aggregations": in effect, people sharing similar genetically and neurobiologically based aptitudes. Thus, if quantitative genetic differences underlie ethnic divisions, qualitative genetic differences underlie class distinctions, to the extent that they are based on achievement rather than ascription. These relationships are summarized in Table 2.1

Finally, a mentalist perspective permits the possibility that people may value their friendships more than their institutional or kinship ties. Friendship may transcend economic interest and political loyalty, and should not simply be reduced to or explained by various genetic or cultural factors. Thus Brain (1976, p. 151) in his cross-cultural survey of "business friends," states that, "Commerce and

Table 2.1
Quantitative and Qualitative Genetic Similarity: Their Hypothesized Relationships with Relation to Kinship, Social Class, and Bonding

		Qualitative Genetic Similarity	
		Low	High
Quantitative Genetic Similarity	Low	Absence of a) kinship ties b) similar social class positions c) bonding	a) Absence of kinship ties b) similar social class positions c) bonding based upon informal friendships
	High	a) Presence of formal kinship ties b) absence of similar social class positions c) absence of bonding based upon informal friendships	Presence of a) kinship ties b) similar social class positions c) bonding based upon informal friendships

gift exchange not only bring people into friendly contact, with mutual interest leading to toleration, cooperation, and peaceful alliance, but the values of friendship may override the values of the commercial exchange.'' And E. M. Forster's well known dictum (1950, p. 626), ''If I had to choose between betraying my country and betraying my friend, I hope I should have the guts to betray my country,'' has not failed to find embodiment within history. John Mitchell, H. R. Haldeman, and John Ehrlichman, for instance, placed their personal loyalty to Richard Nixon above any higher sense of responsibility to the American constitution or, for that matter, to their own best interests. And the relationship was reciprocal. ''Nixon's mistake,'' William Safire (1975, p. 658) writes, ''was in putting political loyalty—first to Mitchell, later to Haldeman and Ehrlichman—ahead of his oath to office.''

To sum up, then, we find two major kinds of social attachments in society: (1) those quantitatively based genetically, which may be referred to as kinship bonds; and (2) those qualitatively based genetically (and neurobiologically), which may be further subdivided into mating and social or friendship bonds. These various social ties need not be exclusive and friendship bonds in particular may include, as well as transcend, either kinship or mating bonds, as in the case of identical twins. A spouse's best friend, for another example, may be his mate; but if so, the relationship was not dictated by the marriage vow alone. Friendship bonds, finally, may transcend considerations of economic interest or of political loyalty.

Bonds, Networks, and Organizations

Recently, a biosocial picture of bureaucracy has emerged which sharply contrasts its historical origins, based primarily upon kinship, to its contemporary basis in Weberian rationality (see e.g., Caton 1982). This picture, however, may be misleading in its historical and contemporary formulations. Nepotistic practices were no doubt prevalent in early bureaucracies, just as they persist in more restricted form to the present day (cf. Balch 1986). It may be an error, however, to view nepotism simply as a quantitative reflection of a proportion of genes held in common. A fair amount of genetic, environmental, and social diversity may characterize kin groups; therefore, the basis exists for relationships that range from those that are quite hostile to those that are quite warm. An uncle might favor one nephew out of many, even actively opposing the advancement of another. Brain (1976, p. 16) points out that within the African tribal society of the Bargwa, personal preference may significantly qualify the character of formal kin relationships. As Willhoite (1986) writes concerning political organization within the early state: ''Succession to certain official positions often became hereditary within particular lineages, but ability or favoritism could determine the specific appointee.'' The larger point is that within kin groups exists the basis for the formation of hierarchic and, therefore, social class and status distinctions.

The view of contemporary organization as an impersonal[4] meritocracy may also be overdrawn. Jay (1971, p. 81) writes that when he first reached managerial status he was determined that he ''would always pick the best candidate irrespective of any irrelevant personal feelings'': ''That resolution fell at the first fence. My very first vacancy produced a candidate who by any objective measurement was noticeably better than the others, but I just could not see myself working with him without a harsh grating sensation somewhere in the region of my colon. Nor could I see any of the rest of the team taking to him.''

Jay goes on to observe that objective qualifications alone would be ''all very well if we were appointing solitaries,'' but if a candidate is to join an existing organizational group ''then all the appropriate qualifications in the world will not counterbalance the gut feeling that he will not fit in, that he is not 'our sort' '': ''You can try and describe 'our sort' in objective terms, but in the end you are merely rationalizing the irrational. The only way to tell is by whatever corresponds to the canine sniffing procedure, and if you do not like the smell there is no point in going any further. Nor should you try'' (Jay 1971, pp. 81–82).

According to Livingston (1974, p. 192), ''superior managers'' have difficulty in making their standards explicit ''because their selection process is intuitive and is based on interpersonal intelligence that is difficult to describe'': ''The key seems to be that they are able to identify subordinates with whom they can probably work effectively—men with whom they are compatible and whose body chemistry ['brain chemistry' may be a more apt characterization] agrees with their own.''

This bonding process[5] may not be simply random, arbitrary, and hence dysfunctional for the organization. First, as Jay suggests, if members of the organizational groups are socially incompatible with one another, they can hardly be professionally effective. Churchill (1948, pp. 239–40) observed, for example, that ''no Foreign Secretary can do his work unless he is supported constantly by his chief. . . . To make things go smoothly, there must not only be agreement between them on fundamentals, but also a harmony of outlook and even to some extent of temperament.''

Second, many of the aims and standards of the organization may be incorporated within the basis for the attachments. As Jay (1971, p. 82) asserts, ''mostly 'our sort' is related to the sort of people who have succeeded at the task as members of the group.'' I recall a review of Jencks et al.'s *Inequality* in which the reviewer saw the apparently random encounter of Jencks with Harvard sociologist David Riesman—which led to Jenck's appointment to a Harvard position—as an example of the large role that chance plays in human affairs. Yet the fact that individuals of similar talent and interest such as Riesman and Jencks should travel in similar circles (networks) is not surprising. Their encounter was more reasonable than arbitrary, as were the likely grounds of Riesman's forming a friendship with Jencks. For if the latter had been an unintelligent, uninformed bumpkin (in which case he never would have met Riesman in the first place), no bonding would have occurred. Riesman and Jencks, literally and figuratively, may have been on the same brain wave.

The upward mobility of able individuals may rest on their recognition and encouragement by their seniors and superiors, whose own similar interests and abilities may prompt bonding between them. Zaleznik (1979, pp. 174–76) points out that the success of "people with great talent" may depend upon their forming "a deep attachment to a great teacher or other benevolent person who understands and has the ability to communicate with the gifted individual." Zaleznik concludes that "psychological biographies of gifted people repeatedly demonstrate the important part a mentor plays in developing an individual" and he illustrates this point by referring to the positive relationship Dwight Eisenhower enjoyed under the tutelage of General Fox Conner early in his military career. There is, in short, the occurrence of what Zaleznik refers to as "self-selections" when generations meet and the more gifted and sensitive of the older generation form tutorial bonds with their like-minded juniors and thereby promote their mobility.

None of the foregoing should deny the real possibility of the formation of bonds and networks along restrictive, discriminatory lines. "Old boy" networks, for example, have been both a historical and contemporary reality and their exclusive, discriminatory, and hence dysfunctional character can hardly be denied or justified.

Individuals who, by virtue of similar abilities, interests, and outlooks, are likely to bond together, are also likely to form networks, defined here as an informal web of social ties that comprises numerous individuals sharing mutual interests and extending spatially over territory. The term is not used in either a loose "metaphorical" sense or as a "precisely defined analytic concept" (cf. Barnes 1972, p. 1). Rather, it refers to any empirically observable set of bonds that unify a group of individuals. As Nobel (1973, p. 6) points out, such a concept may "provide a bridge between the sociological analysis of human behavior and the personal or motivational aspect."

Whereas a bond is a close attachment between two people in a given place (or places), a network includes many people who have formed bonds conceivably over a wide area. Furthermore, although an underlying mutual interest may potentially unite any or all of the network members, these members are not necessarily all familiar with one another. Nonetheless, the underlying basis for the network and the existing social contacts can readily facilitate bonding as, for example, when one member moves to another community or looks for another job. Individuals may retain personal bonds even as they leave a larger network, or they may become members of multiple networks. Broder (1980, p. 132) observes, for example, that in "real-world politics," individuals "are often tied into more than one network."

Broder (1980, p. 131) distinguishes between vertical and horizontal networks. The defining characteristic of vertical networks is that membership is inherited: "Membership was acquired by being born in a certain place at a certain time, of a certain family, race or political affiliation" and was "handed down from generation to generation." Against these older, traditional networks, horizontal networks are "shared by people who were brought together, often from widely dispersed locations and diverse backgrounds, by a particular historical crisis or cause" and although the associations developed have a temporal foundation, they "tend to be very strong, and to endure as common bonds for many years." If not overdrawn, this distinction may prove of value.

Horizontal networks, which are our focus, derive their substance from informal relationships that are freely chosen, and they may therefore exist independently of formal organizations. Networks may flourish within formal organizations, but they clearly need not be isomorphic with the organizational membership. More than one network, after all, may coexist with others within the same organization. More importantly, a network may transcend a given organization. It may include individuals from any number of organizations along with those who are organizationally unaffiliated. Networks are thus more fluid and amorphous than organizations.

Networks, however, may develop within organizations as well as outside them. As Ayres (1983, p. 68) recently observes concerning the U.S. Foreign Service: "There is still an old-boy network within the State Department, but it is now based more on friendships made in the service than at some upper-crust university or mahogany-paneled men's club."

Once a network is formed, whether in or prior to governmental service, it may persist following its governmental years. Thus, one may note Broder's observation (1980, p. 130) that certain networks "will always be, in a certain sense, 'Kennedy People,' or 'Nixon People,' or 'Carter People.' " A member of a network may accordingly move from one organization to another and still remain within the network. In this sense, the network is prior to and more basic than the organization.

From an alternative standpoint, however, it is the filling of formal roles that confers power and status. A recognition of the informal sphere of influence of the network—an influence in no way diminished, but perhaps even heightened by its informal foundations—graphically changes such a perspective. Heclo (1978, p. 102), for example, describes the emerging influence of what he terms "issue networks" in American politics and contrasts his position with those that stress the power associated with more formal structures:

> Iron triangles and subgovernments suggest a stable set of participants coalesced to control fairly narrow public programs which are in the direct economic interest of each party to the alliance. Issue networks are almost the reverse image in each respect. Participants move in and out of the networks constantly. Rather than groups united in dominance over a program, no one, as far as one can tell, is in control of the policies and issues. Any direct material interest is often secondary to intellectual or emotional commitment. Network members reinforce each other's sense of issues as their interests, rather than (as standard political or economic models would have it) interests defining positions on issues.

The political and social domains may, however, influence one another in a reciprocal manner. Civil associations, Tocqueville (vol. II, 1955, p. 123) notes, "facilitates political associations but, on the other hand, political association singularly strengthens and improves associations for civil purposes." (Although Tocqueville is clearly referring to formal groups, he is also quite mindful of their informal basis: cf. vol. I, 1955, ch. xii.) Heclo's "issue networks," after all, owe much of their development to the growth of the national government. Thus we have the following reciprocal relationship: First of all, in Tocqueville's words (1955, vol. II, p. 123), "Politics gives birth not only to numerous associations, but to associations of great extent." The New Deal-Great Society programs have in particular spawned new associations. Writing during the Carter years, Heclo (1978, pp. 105–106) states: "The many new policy commitments of the last twenty years have brought about a play of influence that is many-stranded and loose. . . . More than ever, policy making is becoming an intramural activity among expert issue-watchers, their networks, and their networks of networks."

Heclo is thus emphasizing the importance of the informal self-selecting activity of individuals independently of their formal status:

> Powerful interest groups can be found represented in networks but so too can individuals in and out of government who have a reputation for being knowledgeable. Particular professions may be prominent, but the true experts in the networks are those who are issue-skilled (that is, well informed about the ins and outs of a particular policy debate) regardless of formal professional training. More than mere technical experts, networks people are policy activists who know each other through the issues. Those who emerge to positions of wider leadership are policy politicians—experts in using experts, victuallers of knowledge in a world hungry for right decisions. (1978, pp. 102–03)

In order to see more fully how the informal network may relate to the formal organization, we will now consider the concepts of the organizational psychostructure and the cluster.

Psychostructures: Clusters and Organizations

One of the processes of social selection that involves organizations, according to Maccoby (1976, p. 173), pertains to the selection of "character types" from "the available population to man these organizations" (cf. Kaufman 1960, ch. 6). Maccoby is describing what he calls the "psychostructure" of the organization "that selects and molds character."

We will consider how the organization through its psychostructure may stimulate and reinforce some psychological traits as opposed to others in the final section, but a brief comment should be made regarding Maccoby's contention (1976, pp. 173–74) that "the concept of a psychostructure which selects character types for different roles resolves an argument between the sociological minded, who claim that the role determines behavior, and the psychologically minded, who claim that personality determines behavior." Aside from the fact that the Frommian concept of a "character type" is too typologically formulated, the first point to be made is that Maccoby's environmentalist perspective renders the individual in either case a passive recipient of external pressures. For "character is formed first in family and school" (Maccoby 1976, p. 176) and therefore the self-selecting process at work lacks any underlying genetic dimension as well as any real autonomy on the part of the individual. In the analysis to follow, the latter two dimensions will also be included within the concept of the psychostructure.

With this enlarged notion of the organizational psychostructure in mind, we can consider the concept of the cluster. Within the framework of cultural microevolution, Mead (1964, pp. 154–55; emphasis in the original) uses this concept as the "significant *biological* unit," one which involves "the specific genetic constitution of the members of each cluster of particular interacting individuals."[6]

By stressing the cluster as the unit of cultural evolution, Mead avoids an undue emphasis on the Great Man in History, as well as a complete disregard for the influence of the individual. The cluster mediates between the individual leader and the larger society; it may refer, Mead points out, to the members of the British Cabinet, a scientific elite, or a select intellectual circle such as Bloomsbury. In each of these cases, a cluster of individuals centers about an individual or a small group of individuals of unusual talent or high position. It is principally this fact that appears to distinguish the cluster from the more inclusive network. Thus, Bloomsbury constitutes not just any network, but one that was extraordinarily talented and historically significant.

Here is how Mead (1964, pp. 176–76), for purposes of political analysis, would apply the concept to a small village in a folk society:

> In a village of 300 or 400 people, if one is making a study of interactions and response, leadership and followship, and specialization of role—the formal role of the hereditary magician or the group-selected peace chief or the informal role of the self-selected shaman—it is necessary to ask not only whether a given individual is intelligent or stupid and what his life history has been, but also how much more intelligent or more stupid he is than the other members of the group. It is necessary to take into account both the special capabilities, skills, and so on, and the formal position of each man with these properties.

Mead (1966, pp. 182–86), in fully acknowledging the unique and indispensable role that genius may play in human affairs, also argues that "a devoted and skillful follower may enable a gifted individual to realize a potential that he could never have realized alone." Consequently, it becomes necessary to study the cluster of individuals that the person of extraordinary talent attracts and selects. Historically, Mead points out (1964, p. 187), such study may be difficult because there are lacking "enough specific details of individual interaction" in such cases as those of the American Founding Fathers; Napoleon, as First Consul, working with the *Conseil d'Etat* on the formulation of a uniform civil code for France; or the early Russian Bolsheviks in the development of their revolutionary strategy.

In placing the cluster within an organizational context, we can, however, refer to its more recent and better documented manifestations, by focusing on political clusters that have formed about recent American presidents.[7] When describing "the circles" or the clusters surrounding the five presidents from 1960 through 1980, Broder (1980, chs. 2–4, p. 130) concludes:

Each of these circles forms a self-conscious network, based on a set of shared experiences, triumphs, tragedies and misadventures, especially vivid to those who were there. If past experience is any guide, those shared experiences in the early political lives of these young people will remain a strong bond among them as long as they are active. To one another, and to others outside their circle, they will always be, in a certain sense, ''Kennedy people,'' or ''Nixon people,'' or ''Carter people.''

Let us focus on the Nixon Presidency.

THE PSYCHOSTRUCTURE OF THE NIXON CLUSTER

> I had the feeling—and I think in retrospect I'd probably do it again—that I should stand by my friends.
>
> —Richard Nixon, on the Watergate cover up, ''60 Minutes,'' April 8, 1984.

> And chemistry plays a far more crucial role in politics than is usually imagined.
>
> —Zbigniew Brzezinski, *Power and Principle*

The Leader and the Organizational Cluster

Emerson may have exaggerated when he wrote in ''Self-Reliance'' that ''an institution is the lengthened shadow of one man,'' but the actions of the leader surely set the tone for any organization through his key appointments. Machiavelli observed in *The Prince*: ''The first impression that one gets of a ruler and of his brains is from seeing the men that he has about him.'' Or as Safire (1975, p. 9) puts it, ''The company a President keeps reveals a great deal about himself, his political techniques, and his Presidency.''

A leader's key appointments must to some degree reflect his or her need for congenial companionship and, therefore, the peculiar needs of his or her own personality. A leader, first of all, is not exempt from the universal human desire for friends. Francis Bacon in his essay ''Of Friendship'' remarks on ''how high a rate great kings and monarchs'' set on the ''fruit of friendship''—even ''at the hazard of their own safety and greatness,'' and even to the point of raising their servants to become ''companions, and almost equals to themselves.''

Although certainly not a servant, John Mitchell—a state- and municipal-bond lawyer and one of Nixon's law partners in New York in the mid 1960s—was not a prominent national figure when Nixon made him his campaign director in 1968 and then Attorney General in his administration. Of his intimacy and influence with the president, however, there is no doubt. ''I counted him my most trusted friend and advisor,'' Nixon (1978, p. 339) was later to write, ''and I wanted to have his advice available on the whole range of Presidential decision making.'' As Richard Whalen, a former Nixon aide, maintains, ''Mitchell was Number 1, tied to the White House by a direct line, the uniquely intimate counselor to whom Nixon turned on every subject from minor political matters to Supreme Court appointments'' (cited in Lukas 1976, p. 4). Even when an important appointment goes to a professional and not a friend, a congenial relationship may be necessary for it to be an effective one.

Of course, a relationship that is initially professional may develop into a friendly one. Churchill (1949, p. 21) wrote concerning his relationship with the British Chiefs of Staff during World War II that ''we came very close together indeed, and friendships were formed which I believe were deeply valued.'' Of those who stayed on as his subordinates at the State Department during his term as secretary, Kissinger (1982, p. 440) reports, ''Many became close personal friends.'' When H. R. Haldeman first met Nixon in 1951, there were, according to Rather and Gates (1974, pp. 121–22), ''no immediate vibes or intuitive meeting of mind,'' but Haldeman proceeded to join the Nixon campaign staff and to bring ''to his campaign labors'' the ''same managerial skills that were making him a success in advertising: crisp organization, precise attention to detail, and a strong sense of image, or media.'' But the compatibility of their personalities was clear by his administrative chief of staff. ''There is the man,'' Safire (1975, p. 11; emphasis in the original) writes, ''whose relationship with his President was similar in a way to that of Cathy and Heathcliff in *Wuthering Heights*—Cathy insisted she was not so much in love with Heathcliff, as she *was* Heathcliff—so too did Haldeman see his identity merged with Nixon's.''

Safire (1975, p. 10) characterizes the Nixon-Kissinger relationship as follows: ''Henry Kissinger is shown here to be an extension of Nixon's mind, and a fierce reinforcer of Nixon's penchant for secrecy, since the 'bold strike' and the leap frog technique were the essence of the Nixon way of working. 'Henry plays a hard game,' Nixon once said, with admiration in his voice, because that is the kind of game Nixon liked to play himself, most often through others.''

One appointee who especially reflected Nixon's combative nature was Charles Colson. ''With his own hard-hat eagerness for combat,'' Lukas (1976, p. 12) writes, ''Colson quickly became the point man for the President's search-and-destroy missions . . . there was just about nothing Chuck Colson wouldn't do for one of his heroes. 'Colson could do anything,' Nixon once said, 'He's got the balls of a brass monkey.' ''

The Leader's Lieutenants and Organizational Subclusters

In tapping his top lieutenants, Nixon also tapped into the networks of which they were a part and which in many cases they had formed. Thus, Haldeman had cultivated ties from his college years at UCLA (e.g., with John Ehrlichman) and, later, as advertising executive at J. Walter Thompson, had developed a further coterie:

> Through the early and middle sixties, he recruited a corps of earnest young junior-executive types. . . . In attitude and appearance, Haldeman's proteges seemed to come from the fifties, young men who still adhered to the conventional and quiescent values of the 'silent

generation.' This being the case, it is hardly surprising that they shared his enthusiasm for Richard Nixon. . . . And in time no fewer than five of these well-scrubbed straight arrows—Ronald Zeigler, Dwight Chapin, Larry Higby, Ken Cole, and Bruce Kehrli—would move on, with Haldeman, to Washington and the Nixon White House. (Rather and Gates 1974, pp. 131–32)

Different and competing networks formed about other key Nixon appointees. Safire (1975, p. 99) observes that "in the Presidency, a President should—and does—keep all kinds of company necessary to govern"; therefore, his "palace guard" includes "men of conflicting characters." As Safire goes on to say, "They do not always agree with each other, but Nixon does not always agree with himself."

The Nixon cluster, in short, breaks up into the various subclusters that formed about his top administrators—Haldeman, Mitchell, and Kissinger being prime examples. The subclusters are a product of the leaders' lieutenants in the same way that the cluster as a whole reflects the preferences of the leader. Haldeman (1978, p. 53; emphasis in the original), for example, writes that his wife, "getting glimpses of my operation and relationships with the staff, especially the junior men, very perceptively has since observed that it wasn't *Nixon's* character or moods that allowed Watergate to happen," pointing out that "*my* character and demands pushed people to become 'little generals.' "

Next to the leader, then, the lieutenant determines the organizational psychostructure.[8] Aside from the likelihood that any leader will have a complex and multifaceted personality, which will be reflected in his or her personnel choice, genetic diversity alone will dictate differences in his or her key appointees. In sum, no organizational psychostructure and, therefore, no organization can be completely monolithic and cohesive. There must be subclusters within the organizational cluster as a whole that will be distinct from, and probably competitive with, one another.

The Lieutenant's Lieutenant and the Clique

Organizational diversity and fragmentation are carried one step further by multiple formations of that most cohesive and exclusive of social units, the clique. The clique is as monolithic a social grouping as it is possible to form. According to Jay (1971, p. 37), who refers to it as a "ten-group," based on its likely size, the clique constitutes "the foundation of every corporation" and presumably of every organization. Jay speculates that small cohesive groups of about ten members composed the optimum hunting band during our prehistoric past, thus providing an evolutionary basis for contemporary organizational life, insofar as it continues to favor intimate social interaction (cf. Emmert 1986).

Self-selecting processes are especially likely to affect the composition of the clique. Mutual congeniality is a critical factor for both the clique and the prospective member.

Cliques may exist at any level of an organization, but in this analysis they are viewed as possible components within the subcluster. Two cliques that formed within the Nixon administration are mentioned for purposes of illustration. One formed about Haldeman's protégé, Dwight Chapin, whose college days at USC included fraternity ties with Gordon Strachan and Donald Segretti (Segretti was later chosen by Chapin and Strachan to engage in a "dirty-tricks" campaign operation during the 1972 election).

The other clique centered about the friend (Howard Hunt) that Charles Colson picked during the summer of 1971 to plug the unauthorized leaks the Nixon administration was trying to stop. Hunt was soon instructed by John Ehrlichman to work with his aide and protégé, Egil "Bud" Krogh, Jr., whose "skill" and "decisiveness" impressed Hunt (Lukas 1976, p. 87). Another new man recommended by Krogh and hired by Ehrlichman was G. Gordon Liddy, whom Hunt liked "immediately because he was 'decisive and action-oriented, impatient with paperwork and incubations of bureaucracy.' " "We clicked immediately" and "became fast friends," Liddy (1980, pp. 204, 216) was later to write. In view of recent American history, we can see how a bureaucratic clique may reflect the ultimate wishes of its leadership and at the same time translate them into practical action in its own autocephalous manner.

CONCLUSION

Networks and their social bonds may independently influence organizational personnel selection and thereby the organization itself. As seen in the case of the Nixon cluster, the choice of H. R. Haldeman as presidential chief of staff also meant bringing in no fewer than five of his J. Walter Thompson protégés, not to mention a college friend from UCLA, John Ehrlichman, to the administration. The subclusters and cliques that resulted from this and other key appointments were to have their impact during the Nixon presidency.

Network ties and the values and sentiments that underlie them may have their own independent status and impact. As Hutter (1978, pp. 3, 4) points out, those who would completely explain friendship on the basis of instrumental exchange and reciprocity relationships ignore "the subjective meaning content of a friendship which seems to be unusually important." Personal loyalties can run deep and may transcend other economic and political considerations.

The bonds within the Nixon cluster could be intense. Dwight Chapin, who worked initially for Haldeman at J. Walter Thompson and subsequently for Nixon (in 1966) as his personal aide, was described by Lukas (1976, p. 152) as a "super-loyalist" for "the man he was sure would become the greatest President in history" and one who "remained as loyal to Haldeman as to Nixon."

For his part, Nixon tried to reciprocate. T. H. White (1975, p. 202) writes: "The President balances the loyalty of his men against their blunders and his heart tugs him to reciprocal loyalty. All through the transcripts there is this one admirable combat quality in Nixon—he wants to stand by the men who stand by him. He will push loyalty to the furthest until, when he withdraws it, he withdraws it only to save himself.'' Clearly social bonds by themselves exerted their own impact within the course of the Nixon presidency.

The environment of the organizational cluster reflects and possibly magnifies the underlying genetic traits of the individuals who make it up and this synergistic development may have its own political repercussions. If heredity as well as environment influences nearly every trait and if individuals bond on the basis of similar traits, it follows that the networks and clusters they form will take on similarities that are in part genetically influenced.

Such a perspective qualifies the environmentalist emphasis currently found within the social sciences. Maccoby (1976, pp. 175–77), for example, in observing that corporations tend to strengthen those individual traits that are most useful for corporate work, views the process in predominantly environmentalist terms: ''Most young people are a mixture of still-malleable attitudes. Upon entering the corporation, they still have the chance to become more idealistic and just or they could become disillusioned and self-serving. The traits stimulated in the corporations will in many cases have a decisive effect on the kind of people they become, not only as managers, but as citizens, husbands, wives, fathers, and mothers.'' Maccoby obviously minimizes any self-selecting influences emanating from the incoming employees.

Similarly, Janis understands what he refers to as ''groupthink'' in largely environmentalist terms. Janis (1972, p. 9) defines *groupthink* as ''a mode of thinking that people engage in when they are deeply involved in a cohesive in-group, when the members' strivings for unanimity override their motivation to realistically appraise alternative courses of action.'' Janis's perspective is wide; he sees his work as being ''at the intersection of three disciplines—social psychology, political science, and history'' (1972, p. vi); but as one can plainly see, biology is not one of the three disciplines.

Janis's environmentalism is evident in his brief theoretical introduction in which he stresses previous research on small groups that highlight the pressures on members to conform to group norms. Thus ''the more cohesive the group and the more relevant the issue to the goals of the group, the greater is the inclination of the members to reject a nonconformist'' (1972, p. 5).

But what accounts for the cohesiveness of the group and the salience of certain issues and goals? Janis (1972, p. 192) does indicate a sensitivity to the possibility of a self-selection process at work in promoting groupthink: ''Richard Barnett, in *The Economy of Death*, emphasizes the homogeneous social and educational backgrounds of the officials who man the top posts in Washington. This type of homogeneity may also be an important factor that increases the chances of groupthink.''

Social and educational backgrounds, however, are presumably products of the external environment, or at least they are implicitly so interpreted here. But Janis immediately adds, ''Groups of individuals showing a preponderance of certain personality and social attributes may prove to be the ones that succumb most readily to groupthink.'' Personality attributes, in other words, might be susceptible to explanations that include genetic factors, but Janis does not raise this possibility. Indeed, Janis (1972, pp. 202–03) ultimately resists the formulation of a theoretical explanation for groupthink, and his speculations are clearly environmentalist in tone.

Janis (1989, pp. 56–63) extends his environmentalist analysis of groupthink (and its dangers) to business management in general and to Japanese management in particular (pp. 61–63). But it interesting here that Nakane (cited in Craig 1970, p. 10) feels that in Japan ''the members of a given group actually come to resemble one another,'' not only ''because of the demand for conformity,'' but also ''because those who do not fit in with the group will drop out, and because a given group will attract similar personality types.'' Note the implicit recognition of a self-selection process.

The possibility of a genetically influenced self-selection process as one factor underlying the groupthink phenomenon can be seen with a thought experiment. Recall that the social ties between identical twins are unusually close. Now suppose an organization were made up of the same individual cloned a thousand-fold. Would not that organization turn out to be highly cohesive and committed to the same goals? If so, would not the organizational environment be reflecting the underlying genetic uniformity? Indeed, would it not magnify the genetically based tendencies? Wilson's concept of a ''hypertrophy'' (1979, p. 97), ''the extreme growth of pre-existing structures,'' would seem to fit this phenomenon. If, as during the Nixon years, the top governmental echelons are occupied by networks of individuals with unusually fierce personal loyalties, will not the nature of the government be affected?[9]

The ''synergy,'' to use Corning's term (1983), that results when like-minded individuals cluster serves to multiply the intensity of attitudes held in common, so that networks that feed like-minded individuals into key governmental posts will yield such synergistic effects.

People as well as formal positions must be studied in organizational clusters. Mead (1964, p. 176) makes this point when she concludes: ''A stupid man who holds the position of a hereditary chief will play a different role from a stupid man with no claims to rank. So at all points it is necessary to analyze the formal structure, which prescribes whether descent, primogeniture, a physical property like strength, or a cultivable quality like oratorical ability will determine leadership positions, in relation to the actual constitution-temperament and character of the individuals who do or might fill each of these prescribed positions.''

It is also clear from Mead's remarks that it makes a difference whether it is a smart or a stupid man who holds the position of hereditary chief. The formal position as such does not tell you the personal characteristics of its holder, so these must be determined by the observer. To avoid or minimize the role of the unique properties of individual role-holders because they do not lend themselves to the development of a nomothetic science is to insist on an account that will be as sterile as it may be precise or elegant. The problem of overgeneralizing can be seen in

Janis's implicit application of the symptoms of groupthink to Watergate (1973). These symptoms were identified by generalizing on the basis of an examination of the decisions made by the American government in four crisis situations: the immediate pre-Pearl Harbor situation, the response to the invasion of North Korea, the Bay of Pigs, and the escalation of the Vietnam War.

Safire (1975, ch. 5) finds some of the applications to Watergate problematic. Janis's first symptom, for example, reads: "An illusion of invulnerability, shared by most of all the members, which encourages excessive optimism and encourages taking extreme risks." To which Safire (1975, p. 273) responds, "No, if anything, the Nixon men were afflicted by an illusion of vulnerability."

Safire proceeds to qualify a number of the other symptoms applied to Watergate, but the overriding point is that for any organizational cluster, peculiar environmental, personal,[10] and genetic factors will always co-exist in some significantly unique constellation. And that means that the organizational cluster must be viewed, on the one hand, within its special historical setting and, on the other, in the light of the unique individuals who make it up.

NOTES

1. Indeed, identical twins may be too close to one another for their own good, especially if they are treated by their parents as if they were entirely alike and so dressed alike, for instance. In that event, they are likely to lack any firm sense of self-identity. Thus Banks and Kahn (1982, pp. 41–46) discuss "excessively dependent twins" and their "excessive merging and mirroring personalities." The Goertzels (1978, p. 327) note in their study of 300 prominent personalities the underrepresentation of identical (as well as fraternal) twins in their sample, and they use the life of Eleanor McGovern to illustrate the problems to be overcome by an identical twin who experienced excessive closeness in early life (p. 328).

2. Individuals who tend to travel in the same circles may nonetheless share a natural or intuitive dislike of one another—the reverse of a natural attraction. Thus Brzezinski (1983, p. 24) notes that President Carter "simply did not get along well with" Senate Minority Leader Howard Baker: "The chemistry just was not there." Weinstein (1978, pp. 16, 17), after citing Richard Nixon's personal response to his initial encounter with Alger Hiss (" 'His manner and tone were insulting in the extreme' "), argues that "this negative and emotional response to Hiss contributed to Nixon's decision to keep the [1948 Hiss perjury] inquiry alive."

The presence and political significance of such feuds is set forth by Safire (1985, p. 23) as follows: "In Israel no love is lost between Shimon Peres and Yitzhak Rabin. . . . In the Kremlin, blood feuds rage among the party, KGB and army factions—as well as within each faction, as the Gorbachev-Oqarkov split demonstrates. Surely these are not impersonal 'policy disputes'.

"At home, the clashes between . . . George Shultz and Caspar Weinberger (among others) . . . are not merely outgrowths of differences in approach or conflicting ambitions, but are affected by the element beyond the 92 found in nature and 16 made by men: personal chemistry's human element."

3. This admittedly is a broader definition than that found within the literature. Hinde (1982, pp. 64–65), for example, confines the term *attachment* to the affectional bond formed between mother and infant or to those aspects of adult interpersonal relationships in which one of the individuals is perceived as being stronger and/or wiser than the other. For our purposes, more egalitarian relations should be subsumed within the compass of the terms *bond* and *attachment*.

4. Bell (1981, p. 86) declares that "bureaucracies are based on total impersonality." Emmert (1986) argues, however, that "the seemingly insatiable desire of bureaucrats to hold meetings when a phone call or memorandum would perform the task more efficiently may partly stem from our inclinations for more intimate interaction," which stem from our evolutionary past.

5. Jay explicitly borrows the concept of the bond from the work of Tiger (1969), who sets it within an evolutionary framework. Tiger asserts that during prehistoric times males found it necessary to form close bonds in hunting parties for purposes of survival and that such male bonding persists to this day and indirectly influences, if not determines the governance of society. We do not attempt to judge the validity of Tiger's specific theory (but cf. Brain 1976, ch. 9), but we accept the more general formulation that all forms of human bonding have an evolutionary basis. (Bell [1981, p. 76], rejects Tiger's work in part because "friendship patterns are learned behavior for both men and women"; but surely they are also influenced and limited by genetic and biological factors.)

6. The recent controversy signalled by the publication of Freeman (1983) over Mead's environmentalist outlook in her first work, *Coming of Age in Samoa* is unfortunate in that it obscures the fact that Mead, following World War II, began to develop an explicitly biocultural approach of which the cluster concept is a prime illustration (cf. Mead 1964, "Preface").

7. By focusing on presidential clusters, most clusters are eliminated that may form within the lower echelons of the American Civil Service. I do not mean, however, to suggest that the civil service, enveloped and protected by its own merit system, is therefore immune from the factors in this discussion. First of all, the merit system is part of a self-selecting process, albeit impersonally, ensuring—or attempting to—that individuals with similar aptitudes occupy like positions. Second, not everyone who passes a civil service exam actually enters the service—or stays for a long time, i.e., personal factors soon enter the equation and therefore the self-selection process. Third, the actual recruitment of civil servants is in part the function of what Wilson (1983, p. 356) refers to as "the buddy system," whereby civil service officials may favor friends and acquaintances for employment, at times bending, changing, or making the rules to that end.

8. It is of interest that when in January 1985, President Reagan approved the switch of then Chief of Staff James Baker with Secretary of Treasury Donald Regan, each man proceeded to bring with him his chief lieutenants and so gave the organizational change additional impact. One critical factor in the president's approval of the change was apparently his "very close relationship" with Regan: "It's like a natural chemistry," observed Dennis Thomas, a former congressional lobbyist for the Treasury Department who proceeded to perform the same task under the president: "They are political cousins and personal cousins, and the President believes they actually are genetic cousins as well" (cited by Owen Ullman, "Pals from the start: Regan, Reagan, much alike," *The Philadelphia Inquirer*, January 9, 1985, p. 5).

9. Note how this question relates to the formation of strong in-out-group distinctions. From an environmentalist perspective, such sharp distinctions are simply products of group

pressures on the individual. From a sociobiological standpoint, the proportions of genes held in common within a group becomes the critical factor, so that kinship and ethnic groups are especially prone to making these distinctions. Our "qualitative" genetic self-selection analysis does not necessarily deny the foregoing positions, but it enlarges them: In contrast to the sociobiological view, it indicates that unrelated individuals with similar outlooks may cohere in tight-knit groups. Contrary to the environmentalist view, it argues that these similar outlooks are at least in part genetically conditioned. One thing is clear regarding the Nixon administration: "Throughout the White House the 'them versus us,' 'press versus government' psychology was rampant." (Klein 1980, p. 345).

10. There "is one lesson, from my own experience," Nixon wrote in his book *Six Crises*, in a passage applicable to his seventh, "that seems especially clear: reaction and response to crisis is uniquely personal in the sense that it depends on what the individual brings to bear on the situation—his own traits of personality and character, his training, his moral and religious background, his strengths and weaknesses" (p. viii). Certainly H. R. Haldeman and John Dean responded differently to the surge of events during Watergate. If one insists on pointing to their differing circumstances as a prime reason, a valid consideration simply becomes added to, without detracting from, the possible significance of individual differences in personality.

Chapter 3

Shadow Networks and Self-Selection: The Neuropolitics of Covert Action

> A lot of adventurous young Americans were attracted into OSS and they found within its ranks an opportunity to develop those qualities that, unfortunately, seem to be best stimulated by war: initiative, enterprise, daring and self-reliance. But it also attracted psychopathic characters because it offered fantasy, sensation, intrigue, the lure of being a mysterious man on a mysterious mission.
>
> —Phillip Knightley, *The Second Oldest Profession*

> Personality, intelligence, and psychopathologies of various sorts each involve some genetic inheritance.
>
> —Wilson and Herrnstein, *Crime and Human Nature*

> One of the striking things about the Iran-Contra hearings was how much all the alleged conspirators resembled one another in manner, posture, and speech. . . . This doppelgänger phenomenon is by no means an accident and by no means confined to the White House . . . because by and large people tend to select people to work with them who are cut from the same goods.
>
> —Warren Bennis, *Why Leaders Can't Lead*

There may be an intrinsic appeal to covert action with important consequences: People who are action-oriented may be especially attracted to cover action, and their self-selection leads to the critical formation of networks that are based in part on this neurobiologically influenced quality and that serve to reinforce and magnify it. In essence, the thesis of this chapter is that for the covert action network internal biology and external environment complement one another with potentially important political consequences.

CONDITIONS FOR COVERT ACTION

In outline, this chapter proposes that governmental covert action is likely to result if the following four conditions are met:

1. The self-selection of individuals with certain traits, in particular that of an action-orientation. A key assumption made here is that heredity as well as environment influences such traits. Thus, Masters's "first requisite for a rigorously scientific approach to human nature" is followed, which is a "willingness to abandon the belief that answers are either/or: our behavior can be both innate and acquired" (Masters 1989, p. 1; see White 1972).

A further assumption is that individuals, rather than being uniformly and passively molded by their environments, actively seek out congenial environments based upon genetically based propensities. This process of self-selection will be elaborated with reference to action-oriented individuals, where it should be especially applicable (see Scarr and McCartney 1983).

2. The formation of bonds and networks based on self-selection of individuals sharing critical qualities. This process will presumably follow the more general theoretical strictures just described in chapter 2 governing bond and network formation; that is to say, individuals who share similar genetically and neurobiologically influenced interests and orientations will tend to be attracted to one another.

3. A suitable environment is present that includes (a) encouragement from the top and (b) opportunity at the bottom. Private or extra-legal covert action may nonetheless still be possible in the absence of this dual condition. This possibility is elaborated on in the concluding section.

4. The resulting dynamic interaction of the first three conditions characterized by what Corning (1983) terms *synergy* or Wilson (1978) *hypertrophy*. In Wilson's language, hypertrophy involves the social reinforcement of pre-existing tendencies —in such cases, the social environment and genetic tendencies reinforce one another.

Presumably the history of covert action throughout the world illustrates this model. But for purposes of concrete illustration, the recent American examples of Watergate and Iran-Contra seem especially appropriate, not least for the amount of documentation that is available in each case. Both involved illegal activity whose discovery caused scandal, though Watergate was a covert operation turned inward, that is, within the American domestic political arena. The individual careers of two of the most central figures in each instance—E. Howard Hunt and Oliver North—will serve as focal points.

Given a populational perspective, individuals are likely to vary widely in their degree of action orientation. Also referred to as sensation or thrill seeking or as risk-taking behavior (see MacDonald 1988a), this trait involves the degree of environmental stimulation that an individual seeks. Some people seemingly require and seek out environments that promise an intense level of excitement, whereas others deliberately avoid such environments.

Recent research strongly indicates a neurobiological basis for this trait. Action-orientation appears to be related to "variation in the reactivity of the central nervous system and the autonomic nervous system as well as somatic sources of reactivity." Neurobiological factors also include the septohippocampal system, an increased level of gonadal hormones, and low levels of enzyme monoamine oxidase (MAO) (MacDonald 1988a, pp. 58–59).

Action-oriented individuals tend to be disproportionately male (MacDonald 1988a, 229–33). From childhood, males tend to engage in more rough and tumble play than females, to take more chances, and to shun more routine and sheltered activity. These generalizations, of course, do not rule out a wide range of variation within each sex, so that some women will be highly action-oriented and some men not so at all.

Aggressive individuals, not surprisingly, tend to be action-oriented—to be impulsive and to seek high levels of environmental stimulation (MacDonald 1988a, pp. 195–97). Therefore aggressivity and action-orientation tend to be linked.

Action-oriented individuals also tend to be more extraverted (MacDonald 1988a, pp. 49–50). Such individuals seek out a higher level of and a more constant exposure to social companionship. Such a relationship appears to support Hoffer's contention (1981, pp. 111, 112) that "teamwork is rare in intellectual or artistic undertakings, but common and almost indispensable among men of action."

Finally, action-oriented individuals are likely to be less morally inhibited. Taking risks, seeking sensation implies upsetting precisely those routine behaviors that encapsulate moral standards. At the extreme, the psychopath, especially the violent one, is prone to be highly sensation seeking. Moreover, recent research suggests that a neurobiological basis exists for such psychopathic behavior (see Meloy 1988; (MacDonald 1988a).

Of course, few action-oriented individuals are psychopathic. But a populational perspective suggests that rather than a typological polarity of a few psychopaths at one extreme, with everyone else uniformly "normal," a wide range in such behavior should be expected. As Wilson and Herrnstein (1985, p. 207) assert:

> Instead of a sharply defined entity, psychopathy is made up of deviations along some of the common dimensions of personality. . . . Everyone has some level of internal arousal or emotionality and a susceptibility to the conditioning of internalized prohibitions; the psychopath's are relatively weak.

The conventional view is that one's inhibitions, in the form of the conscience or Freudian superego are wholly a societal artifact, that is, environmentally formed. But the fact that brain-damaged individuals or those suffering from such neurologically based disorders as Tourette's syndrome may become quite uninhibited (see Sachs 1985) argues for an important biological dimension underlying moral behavior.

Indeed, the stages of moral development as formulated by Kohlberg (1969) and qualified by Gilligan (1982) implies the active influence of a number of biologically based factors: age, sex, and intelligence. If maturation represents physiological change as well as experience, clearly there is a critical biological underpinning to the differential capacities of children at varying ages to formulate ethical judgments.

Furthermore, Gilligan suggests that, in general, females are more likely than males to embed their ethical judgments within the matrix of concrete social relationships, especially those of family and friends. It is at least plausible that biological as well as environmental factors are at work here. It would be interesting to learn whether the relatively lower level of female action orientation introduces more inhibitions of their behavior. According to Carrol (1989, p. 314), "Women have exhibited what some have labeled greater 'moralism' than men." Put negatively, action-oriented men (and women) may be more ruthless (and thus less caring) in the pursuit of their ends.

Finally, mental capacity may be a factor. Both Kohlberg and Gilligan (1971) indicate this likelihood in relating moral development to Piagetian intellectual development. This linkage suggests that intellectual capacity may be related to moral choice—as perhaps a necessary, if not sufficient condition for a certain level of moral behavior (cf. Wilson and Herrnstein 1985, p. 159).

The general proposition that biological factors influence moral choice has certainly been entertained (cf. Waddington, 1961, E. O. Wilson 1978, and MacDonald 1988b). Typically, this formulation is drawn within the context of human evolution, although couched in typological language, that is, as applying more or less uniformly throughout the species. Our populational perspective would, assuming a biological dimension, imply a range of behavioral inhibitions constraining the moral choices of individuals: Given comparable levels of moral socialization, individuals will respond differently, based in part on a neurobiologically based susceptibility to behavioral inhibitions. At one end of a wide continuum of response is the psychopath, with no inhibitions and no sense of remorse for antisocial acts. At the other end, in theory at least, is the exceptionally inhibited individual, one wracked by guilt for the least perceived social transgression.

To introduce a genetic and neurobiological component does not negate the presence and significance of environmental ones. It simply means that environmental factors do not entirely and uniformly determine individual responses. It may mean, for example, that the most strict moral regimen will meet with limited success in reforming a psychopath, while psychotherapy designed to reduce inhibitions in severely repressed individuals will not wholly succeed. But it does not mean that moral training has no effect; most humans may be indeed by nature susceptible to its influence and without its actual presence would succumb to anarchic behavior in the worst sense.

This biopsychological formulation is admittedly highly speculative, but research is now accumulating to support the critical assumption made here: Individuals vary in behavioral inhibition and such variation has a neurobiological underpinning (see Trasler 1987). As Wilson and Herrnstein (1985, p. 209) conclude: "Personality, intelligence, and psychopathologies of various sorts each involve some genetic inheritance."

The introduction of a genetic dimension to behavior does not rule out the autonomy and significance of mental states of mind. As the title of Katz's book, *Seductions of Crime: The Moral and Sensual Attractions in Doing Evil* (1988) indicates, the phenomenology of criminal attraction deserves attention. As Katz (1988, ch. 9) notes, material circumstances such as poverty cannot by themselves account for crime; rather, existential "thrills" exist as powerful attractions (a position anticipated by Aristotle in *Politics*, Bk. II, 11, "But want is not the only cause of crimes. Men also commit them simply for the pleasure it gives them, and just to get rid of an unsatisfied desire.") In any event, Katz (1988, "Introduction") counterposes his experiential emphasis to the "conventional" and "academic" emphasis on background factors in crime, whether genetic or environmental in nature. This academic emphasis may be too mechanistic and deterministic and, at the same time, superficial and inadequate, but we do not have to view the objective and subjective dimensions as exclusive but rather as complementary. Indeed, the genetic, social, and mental are inextricably bound: Individuals sharing certain genetically influenced traits bond to form social networks that serve to reinforce and intensify the resulting patterns of thought and action.

Furthermore, the genetic dimension, as it is related to action-orientation, may help account for individual differences. Thus Katz may be right to call attention to the "sneaky thrills, that some shoplifters might get, but he nevertheless fails to treat the obvious possibility that some individuals, even if young and male, might never be tempted by or even especially experience any "sneaky thrill."

One final extended introductory note is in order. The personality trait that will be discussed does not exclusively reside in covert-action participants, so something needs to be said concerning those additional factors that may lead an action-oriented individual toward clandestine intelligence activity rather than elsewhere. The following briefly suggests a few of the factors that may be important.

Social class background. Although poverty may not produce crime, action-oriented children growing up in poor neighborhoods may be especially prone to forming street gangs that lead to street crime (see Quay 1987b).

According to Thompson (1967, pp. 176, 177) the 1960s California motorcycle gang, the Hell's Angels, was "very definitely a lower-class phenomenon," with its typical members being rootless, having since childhood lived "in total mobility," relating "entirely to the present, the moment, the action."

The upper-class activist, as we shall see, is a more likely candidate for clandestine operations.

Intelligence. As Wilson and Herrnstein (1985, ch. 6) point out, intelligence is inversely related to violent crime; that is, violent criminals tend to have lower intelligence. One reason for this, according to the authors, is a limited "time horizon," that is, a relative incapacity to plan for the future and therefore a tendency to live in and for the moment. Action-oriented individuals of lesser intelligence seem to fit this category (see Quay 1987a).

More intelligent action-oriented individuals may also commit criminal or violent acts, depending in part on other factors.

A perception of blocked opportunity. Rather than being simply a product of objective circumstances, a sense of entrapment, as will be shown in chapter 7, is influenced by cognitive capacity—the more cognitively able tend to feel more frustrated. Thus, revolutionary leaders have often hailed from relatively more comfortable middle-class families. Political terrorists, as Rubenstein (1987, p. 67) observes, often draw their ranks from middle-class students who later—especially when a recessed or stagnant economy cannot suitably place college graduates—form a waiting-to-be unleashed ''virtual army of idealistic, ambitious, action-oriented ex-students.''

Violent political action certainly need not be restricted to the left. The first Nazis, according to Merkl (1975, p. 14) were disproportionately middle class, and for the Free Corps and stormtroopers in particular, Joachim Fest (cited in Merkl 1975, p. 139) speaks of their ''extremist adventurism and criminality masquerading as nationalism,'' and their ''[a]ctive unrest, readiness to take risks, belief in force and irresponsibility.''

Specialized skills. Intelligent action-oriented individuals, given the opportunity, will presumably self-select themselves in part on the basis of the particular genetically influenced aptitudes that they possess. Again, to take admittedly extreme examples, in the following passage (cited in Arlacchi 1986, pp. 130–31) a close (anonymous) observer describes a mafioso-entrepreneur:

> These people have amazing vitality. They never stand still, they're never idle. One moment they are busy over some deal, later on they're having lunch with friends, then they visit one of their lovers. Then some ''situation'' comes up which they have to ''control'' . . . they are always on the move, and spend hours traveling from place to place in their cars. . . . The whole thing's done with feverish intensity: never an empty space, never a slack moment. . . .

In the following excerpt, Dennis Levine describes his Drexel partner Ivan Boesky, who was at the center of the 1986 insider trader scandal on Wall Street (cited in Stevens, 1987, p. 123):

> The guy appears to be superhuman. At his office by daybreak—sometimes before that, working through the evening without lunch or cocktails or snacks. All he needs is coffee—sixteen to twenty cups a day. Food and sleep and rest and music—that is for mortals. Not for Ivan Boesky.

It is also of interest that in his discussion of the psychopathic mind, Meloy (1988, p. 24) explicitly cites the Boesky scandal as an example of ''nonviolent predation,'' which he finds has ''much in common with more overtly violent acts of predation.''

These two examples suggest some ways in which exceptionally action-oriented individuals may self-select themselves under varying circumstances or as influenced by other qualities. In other words, these individuals are not simply interchangeable, and our focus on likely candidates for covert action will identify background and personality characteristics that may not be shared by others. Nonetheless, the common thread of action-orientation in the examples does allow some possible overlap and spill over into covert action. Here are a few such indications:

1. On November 19, 1965, Ralph Barger, Jr., president of Hell's Angels, held a press conference in which he read a telegram that he had already sent to President Lyndon Johnson, volunteering the Hell's Angels to serve as a "crack group" of trained guerrillas in "behind the lines duty in Viet Nam" (Thompson 1967, pp. 322–23). Although this offer was never acted on and granted that Barger, by virtue of his position as president of his group, may have been unrepresentative of its members in such traits as intelligence and dominance, it is still of interest that he drew an explicit connection between the gang and a possible clandestine paramilitary role in Vietnam.

2. In 1960, Robert Maheu, an ex-FBI officer turned private investigator and a freelance CIA agent with underworld connections, turned to the Mafia with a proposition to kill Castro for $150,000—after being approached by a former FBI associate in the CIA's Office of Security to arrange such a deal (Ranelagh 1986, pp. 353–58). Interestingly, the network that emerged for this enterprise spanned the FBI, the CIA, and the underworld.

The clandestine operator, according to Marchetti and Marks (1974, p. 253), "is quite often on the brink of the underworld, or even immersed in it, and he frequently turns to the least savory types to achieve his goals. Criminals are useful to him, and are often called upon by him, when he does not want to perform personally some particularly distasteful task or when he does not want to risk any direct agency involvement in his dirty work."

3. At the end of World War II, a number of Nazi and Gestapo officials managed to escape detection by taking advantage of the incipient cold war and worked for the CIA or the CIC (Counter Intelligence Corp in Eucom—the European Command—the military authority in the U.S. zone in Germany immediately following World War II). Klaus Barbie was a prominent example (see Ryan 1984, ch. 9).

These examples indicate that a similar mind-set may, under certain conditions, transfer from one situation to another. Katz (1988, p. 321) theorizes that "there may be a fundamental similarity in the dynamics that people create to seduce themselves toward deviance," and he proceeds to link the "killing of defenseless victims to sustain a career of robberies with the deception of democratic publics to support government-sponsored killings of defenseless foreigners." He further notes the inadequacy of "materialist-background" explanations arising from either the Left or the Right to account for either of these activities—whether poverty or capitalist imperialism (the Left) or an overly liberal judiciary and the need for a tough defense deterrence (the Right). Katz (1988, p. 322) rightly points to the nonrational, nonutilitarian character of the mental activity involved in both ghetto crime and foreign interventions. Cost/benefit analysis will not suffice to account for either phenomenon.

Nor, apparently, can it fully account for such Wall Street activities as takeovers. Lewis (1989, p. 222) asserts: "There are those who would have you think that a great deal of thought and wisdom is invested in each takeover. Not so. Wall

Street's takeover salesmen are not so different from Wall Street's bond salesmen. They spend far more time plotting strategy than they do wondering whether they should do the deals. They basically assume that anything that enables them to get rich must also be good for the world. The embodiment of the takeover market is a high-strung, hyperambitious twenty-six-year old, employed by a large American bank, smiling and dialing for companies.''

As noted elsewhere (White 1982, 1992), the subjective realm—Popper's ''second world''—may have its own autonomy and causal potency. It therefore may need to be treated on its own terms, calling for an emic as well as an etic approach. (The subtitles of two other recent works, one on happiness, the other on love—Csikszentmihalyi's *Flow: The Psychology of Optimal Experience* [1990] and Tennov's *Love and Limerance: The Experience of Being in Love* [1989]—explicitly highlight the experiential quality of each feeling.)

Granted that subjective states of mind may take on a life and causal force of their own, it is not necessary to neglect the ''background'' influences on them, whether genetic or environmental. In the subsequent section, we consider these influences within the context of covert action.

COVERT ACTION: E. HOWARD HUNT AND OLIVER NORTH

> Intelligence is primarily action-oriented, frequently calling for immediate political action to cope with dangers abroad.
>
> —Walter Laqueur, *A World of Secrets*

> He [Edwin P. Wilson] filled out his personal history statement, underwent a battery of medical and psychological tests, scoring well as the kind of adaptive, nonreflective, self-reliant, action-oriented personality type the CIA especially prized.
>
> —Peter Maas, *Manhunt*

> Liddy and I get along well together . . . we're both action-minded and see things the same way.
>
> —E. Howard Hunt, *Undercover*

> I readily admit that I was action-oriented.
>
> —Oliver North, Testimony before the Joint Congressional Committee on Iran-Contra

Before considering E. Howard Hunt and touching on his professional association with the CIA, some qualification of the quotations mentioned, especially as they might relate to the CIA, is in order. It is not difficult to find generalizations concerning spies and the CIA that support our focus. Cooper and Redlinger (1986, p. 36) state that ''a good many spies have exhibited psychopathic or sociopathic tendencies, and these may in some measure be characteristic of the breed.'' Allen

Dulles, director of the CIA during the Eisenhower years, writes (1963, p. 175) that while the motivation of the intelligence officer is "a deeper thing than a mere longing for excitement" and "a sense of adventure," these do play a role and that "Education, talent and the highest security clearances will not make him an intelligence officer if he does not have this motivation."

The critical qualification, however, is that no one single type of CIA employee exists. Thus, Dulles (1963, p. 133) also observes that "the jobs in intelligence are manifold and there is room for many kinds of talent." There is presumably a process of mutual selection involved in the filling of these jobs. It is estimated that only 1 percent of more than 150,000 applicants are eventually accepted in any one year by the CIA. Eighty percent of all applicants will not even receive a job interview, usually for lack of formal education. The physical and the lie detector tests fail many who may have passed the interview. In addition, less than 5 percent of CIA personnel are involved in covert, undercover operations (Ellis 1988, pp. 65–67).

According to Powers (1979, pp. 35–37), there are "three distinct types of personality" within the CIA: "the spy runners, the analysts, and the political operators." Richard Helms, Powers continues, was "a classic embodiment of the first personality, a man with the discretion, restraint, and exactitude traditional among the bureaucrats of espionage, whose work is equally divided between the penetration and the preservation of secrets, and whose ideal style is brisk invisibility."

The second personality is that of the analyst, who if "born rather than merely assigned to the job has a glutton's appetite for paper—newspapers and magazines, steel production statistics." It "is no surprise," Dulles (1963, pp. 177–78) writes, "that people from the academic profession fill many of the analytical jobs," being "more interested in intellectual pursuits than in people, in observation and thought than in action."

The "third personality" is that of the covert operator. Powers (1979, pp. 36–37, 82) includes Frank Wisner, Allen Dulles and Richard Bissell within this category: "All were gregarious, intrigued by possibilities, liked to do things, had three bright ideas a day, shared the optimism of stock market plungers [cf. an Ivan Boesky?], and were convinced that every problem had its handle, and that the CIA could find a way to reach it." Enterprises connected with these operators were "aggressive, enthusiastic, and too often morally careless": "The adventurers thought of the world as being infinitely plastic; they thought they could do anything with funds and a broad okay from the top." Finally, this category "tended to be white Anglo-Saxon patricians from old families with old money, at least in the beginning."

A few qualifying observations are necessary. First, an "old-boys" network involving upper-class WASP families may well have fed personnel into CIA covert operations, especially at the outset. The existence of such a network does not mean that all upper-class offspring are action-oriented and potential CIA operatives. It is plausible that an underlying genetic diversity characterizes any large human population, including any social class; therefore, a certain proportion would tend to be action-oriented. Because of differing aptitudes, suitable professions other than clandestine operations might be selected. In any case, an educated WASP would find this employment more accessible than a ghetto youth or a rootless lower-class California motorcycle gang member.

Finally, it bears reiterating that although psychopathic individuals are likely to be action-oriented, the reverse formulation does not hold. Action-oriented individuals certainly need not be psychopathic. Nonetheless, if we view individuals and their moral inhibitions as within a range rather than a stark dichotomy, then we might suspect that action-oriented individuals would tend to be distributed closer rather than further from the psychopathic extreme. What this may translate as is a willingness to "cut corners," to overlook moral niceties, to take morally dubious steps to reach presumably more lofty ends. Politicians are likely to be action-oriented, although as with any class of people, some more than others. Although a stark dichotomy, Barber's activity-passivity dimension (1972) at least gets at the variation in this trait among political leaders, and it is noteworthy that the two most recent presidents whom Barber labels as active and "negative" in personality (i.e., downbeat, moody) were perhaps most likely to have countenanced morally problematic actions in the course of their careers—Lyndon Johnson and Richard Nixon.

Watergate and E. Howard Hunt

In treating Watergate and Iran-Contra through the careers of two of their most central participants, the framework described in the introductory section will be used.

Self-selection and action-orientation. On September 24, 1973, before the Senate Watergate Committee, Hunt (1974, p. 317) described his personal background as follows:

> I was educated in the public schools of Florida and New York, and in 1940 graduated from Brown University. Six weeks later I volunteered to serve in the armed forces. While destroyer officer on the North Atlantic convoy run before Pearl Harbor, I was injured and medically discharged. Later I volunteered and became an Air Force intelligence officer. In 1944 I volunteered for the Office of Strategic Service, the forerunner of CIA; and was sent to China, where I was engaged in partisan warfare until the end of the war.
>
> In 1959 I joined the Central Intelligence Agency, from which I retired on May 1, 1970 . . . I was engaged in intelligence, covert action and counter-intelligence operations.

Hunt's summary of his personal background and career reveals clearly a pattern of self-selection: He volunteers for military action before Pearl Harbor; he later volunteers for intelligence work for the Air Force and for the OSS before joining the CIA. (Hunt does not mention his second career as a spy novelist, based in part on his personal experiences.) We will now look at his professional contacts and personal friendships within this self-selection process.

Bonds and network formation. Most of the critical turning points in Hunt's career correspond to key professional meetings. In September 1944, Hunt (1974, p. 31) got wind that recruitment was going on for a "hush-hush" organization that "specialized in unorthodox, behind-the-lines warfare." Since "the thought appealed" to him and he discovered the head of the organization—the Office of Strategic Services—to be William J. Donovan, his father's old fellow lawyer and social acquaintance in Buffalo, New York, Hunt enlisted his father's aid in arranging a meeting with Donovan, who, while insisting that Hunt would have to meet the qualifications of his "select, elite force," was "cordial," welcomed Hunt's interest and clearly facilitated his entry in the OSS, which took place almost immediately. Donovan is described by Powers (1979, pp. 21–22) as a "man of enormous crude energy and the open adventurous mind which was to characterize American intelligence until the Bay of Pigs." According to Knightley (1986, p. 225), Donovan "saw himself, above all, as a man of action."

Following the war, as a part of the European Cooperation Administration (ECA), Hunt met (in the summer of 1948) a deputy assistant secretary of war, Frank Wisner, who had also served in the OSS and had favorably impressed Hunt as "a hard-line anti-Communist."

Apparently Hunt's impression of Wisner was reciprocated, for about Easter 1949, Frank Wisner, while in Paris, paid a visit to Hunt. He told Hunt that he knew of Hunt's service in the OSS and that he "was well regarded by mutual friends." He was setting up within the CIA a new organization, the Office of Policy Coordination (OPC), which was to be "America's action arm in the Cold War": Would Hunt "be interested in joining?" Hunt (1974, p. 63) replied that "the idea was attractive," but that his government work with ECA "had not especially appealed" to him, whereupon he was assured by Wisner that Hunt "could work unfettered in OPC." Hunt's twenty-year CIA career was launched, as well as a professional association with Wisner that extended into a social one following Wisner's retirement. Under Wisner, the OPC is described as revelling "only in action": "Rather than coherent, planned growth, there were bursts of wild activity as Wisner plunged off in one or another direction in search of a mission for OPC" (Prados 1986, p. 34).

Following his apparent retirement from the CIA in April 1970, a question still remains as to whether Hunt really terminated his ties; see Lukas (1976, p. 78) and Houghan (1984, pp. 6–9), Hunt was subsequently "bored" with a "routine job in public relations" (Powers 1979, p. 252). According to Lukas (1976, p. 78), he "yearned for some action."

Powers (1979, p. 252) describes the results: "He renewed a casual friendship with Chuck Colson, a fellow graduate of Brown University whose White House connections impressed Hunt. Colson, in turn, was impressed by Hunt's "leading role" in the Bay of Pigs. Lunching together often, they found themselves politically congenial and imagined that their shared taste for intrigue might make them good partners." Needless to say, Colson's role in the Nixon White House was ruthlessly action-oriented. Colson described himself as a "flag-waving, kick-'em-in-the-nuts, anti-press, anti-liberal Nixon fanatic,'' and another White House aide later said, ''If you think what Colson *did* was bad, you should have heard what he *wanted* to do, what he was *bent* on doing'' (cited in Lukas 1976, pp. 11, 12; emphasis in the original).

Hunt proceeded to work for the Nixon White House in July 1971, attempting to discredit three public figures: Senator Edward F. Kennedy, by learning more about Chappaquiddick; former President John F. Kennedy, by linking him directly to the murder of Diem in South Vietnam; and Daniel Ellsberg (who had just leaked the ''Pentagon Papers'' to the press), by suggesting that the CIA prepare an unflattering psychological profile. In addition, in connection with an all-out effort on the part of the Nixon administration to stop unauthorized leaks, spurred in particular by the Ellsberg disclosure, Hunt worked intensively with ''the plumbers,'' the group set up with this goal in mind.

It was in the summer of 1971 that Hunt met fellow-plumber G. Gordon Liddy. As Hunt (1974, p. 156) writes, ''Liddy and I liked each other at once. To me he seemed decisive and action-oriented, impatient with paperwork and the incubations of bureaucracy.'' Liddy (1980, pp. 204, 216) felt similarly: ''We clicked immediately'' and ''became best friends.'' For the next half year the two became ''almost inseparable''; the Hunts were ''the only ones we ever saw socially,'' recalled Frances Liddy (cited in Lukas, 1974, p. 169): ''We would go there for dinner and the men would sit around talking for hours.''

Shadow networks. If personal friendships—bonds—can be formed based in part on mutually held and presumably genetically influenced traits such as action-orientedness, so can a series or network of such bonds. Such a network was formed by Hunt when, in the CIA, he was instructed to assist Cuban exiles in overthrowing Castro. Hunt began recruiting and organizing Cuban Americans for what resulted in the Bay of Pigs. Such clandestine networks may be termed ''shadow networks'' (Marshall, Scott, and Hunter 1987, p. 14). One important proviso is that while Marshall, Scott, and Hunter restrict the usage of this term to the ''right wing,'' there is no reason to exclude the possibility that such left-wing based networks can also exist.

Indeed, Rubenstein (1987, pp. 52–53) faults conservative analysts with overstressing a monolithic, wholly externally controlled (through what was the Soviet bloc) ''terror network'' and for ignoring right-wing networks in connection with terror. If genetically influenced traits might, under certain circumstances, lead to the formation of bonds and networks that lend themselves to clandestine action, no large population should be exempt—and certainly *not* any that generated mass movements, whether on the right or left, that resulted in the establishment of totalitarian regimes.

Shadow networks generated for one cause may disappear but may also remain in a quiescent form that can be reactivated. Hunt (1974, p. 101) notes that ''through friendships and associations I formed in the Cuban project, I was able in 1971 and 1972 to recruit Cuban Americans to assist me in a series of White House-directed operations that came to be known as Watergate.''

It may be a mistake to stress the monolithic, externally controlled nature of any network. Our emphasis on genetic diversity and on the process of self-selection leads us to believe that differences and stresses would characterize any set of personal relationships, but that they nonetheless may be held together largely by mutual personal attraction and the sharing of ends. Conceivably, a CIA-inspired shadow network could operate as a "rogue elephant"; but the likelihood of external influences, especially from higher political circles, cannot be excluded.

Encouragement from the top. "In deciding on covert operations," Treverton (1987, p. 198) concludes, "most presidents have not been captives of organizations they did not understand or processes they could not control. Far from it, more often they put pressure on the CIA rather than vice versa." Treverton (1987, p. 202) points out that presidential encouragement is especially likely when activists in favor of covert operations form part of the president's inner circle, as with Allen Dulles in the planning for the Bay of Pigs (under Eisenhower).

Nixon's own involvement in Watergate ruined his presidency. But earlier that involvement made itself clear to its clandestine operatives. According to Hunt (1974, p. 187), in late 1971, when passing through the receiving line at the White House and telling Nixon that "I'm working with Chuck Colson now, Mr. President, the reply was: "Oh, yes, I know about that." Hunt says that he "was, of course, pleased and encouraged by this knowledge of Presidential awareness of my work for Charles Colson."

Presidential (or upper level) encouragement serves two important functions: First, it provides the opportunity for self-selecting tendencies on the part of the action-oriented to surface. Second, it provides a rationale, a moral fig leaf as it were, for illegal actions that otherwise would be difficult to justify, especially in ethical terms. Although only a very few action-oriented individuals are clearly psychopathic, they are more likely to have fewer behavioral inhibitions and, therefore, require less moral justification for their actions—provided perhaps that there exists at least a minimal moral basis. As Marchetti and Marks (1974, p. 249) put it: "Deeply embedded within the clandestine mentality is the belief that human ethics and social laws have no bearing on covert operations or their practitioners. The intelligence profession, because of its lofty "national security" goals, is free from all moral restrictions." The authors go on to conclude:

> One of the lessons learned from the Watergate experience is the scope of this amorality and its influence on the clandestine mentality. E. Howard Hunt claimed that his participation in the Watergate break-in and the other operations of the plumbers group was "what I believed to be . . . the best interest of my country." In this instance, at least, we can accept Hunt as speaking sincerely. He was merely reflecting an attitude that is shared by most CIA operators when carrying out the order of their superiors.

The resulting synergy. Given the three factors—the self-selection of action-oriented individuals, their formation of bonds and networks, and some encouragement from above—a synergistic process may take hold, intensifying the characteristics that the participants share through their interaction and heightening their subjective awareness in this respect. The Hunt-Liddy partnership is especially instructive. As Houghan (1984, p. 38) writes, "By putting Liddy and Hunt together, the White House ignored the likelihood that the fantasy lives of these two agents would reach a critical mass."

On New Year's Eve 1971, the Liddys and Hunts celebrated together, with the two plumbers withdrawing to draw up an intelligence-gathering plan for Attorney General John Mitchell. Called Gemstone and requiring a budget of one million dollars, this plan included kidnapping squads to capture radical leaders (such as Jerry Rubin and Abbie Hoffman), call girls to compromise Democratic politicians, and mugging squads to rough up hostile political demonstrators. Although these far-out features were rejected and the budget substantially slashed, two points need to be made. First, Gemstone indicates how far the mutual reinforcement of two action-oriented individuals can lead, and second, it sets the tone for future action. Any subsequent activity would appear more moderate in comparison and, hence, legitimate, even though the Watergate break-in was unconstitutional and would result in the removal of a president.

Iran-Contra and Oliver North

Self-selection and action orientation. The major career decisions that Oliver North has made provide a reasonable indication of his interests and aptitudes. His initial career choice was that of teaching, but after his first year in 1962 at a small teachers' college near Rochester, New York, North realized that he wanted "something with a greater sense of immediacy, urgency and importance" and proceeded to enlist in a Marine Corps officer's training program at Camp LeJeune, North Carolina. According to Bradlee (1988, pp. 32–33), North "thrived on it" and decided to make the marines his career by becoming a commissioned officer through the Naval Academy at Annapolis.

Again, North thrived: "He couldn't understand those who had scrawled the initials IHTFP—which stood for "I hate this fucking place"—all over the academy. North *loved* the "fucking place." "The academy basically preached action over reflection, and North gravitated to the philosophy. Above all else he was a doer" (Bradlee 1988, p. 140; emphasis in the original). Boxing and sailing were among his activities, and during holidays when other plebes went home, North traveled to military bases for extra training in such activities as parachuting and survival tactics.

Following graduation and marriage, both in 1968, Second Lieutenant North became a battle commander in Vietnam. There he became close to his company captain, Paul B. Goodwin, who "liked that North was sharp and correct, like him" and that he was "a doer" and "had guts." Later Goodwin asked North to be godfather for his son (Bradlee 1988, p. 71). North won a Bronze Star, a Silver Star, two Purple Hearts and a Navy Commendation Medal, which reported that he carried out "more than seventy day and night ambushes, and devised tactical innovations that resulted in heavy enemy casualties'' (cited in Bradlee, 1988, p. 80).

His intense wartime activity carried over into his peacetime military career. Assigned to head a training facility in Okinawa (separated from his wife), North always participated in every exercise; he never ''just observed'': ''As usual, no one outworked North, and he seemed to revel in the physical intensity of the assignment.'' (Shortly after returning from Okinawa in December 1974, North's wife threatened to leave him, apparently the cause of a mental ''breakdown'' resulting in his hospitalization for twenty-one days).

Selected in 1980 as one of twenty marine officers to attend the naval war college in Newport, Rhode Island, North graduated in June 1981 and was assigned to the National Security Council (NSC) just as the Reagan administration, with its commitment to military strength, was getting under way. Although considered a prestigious assignment, promising opportunity, North's response to the question of whether he wanted it was ''No, absolutely not.'' A desk job, even in the White House, did *not* appeal to him. Ironically, it was his forthright rejection that ensured his selection; according to Robert L. Schweitzer, the military official responsible for the choice: ''I'd rather have someone who wanted to be a soldier or lead troops, than someone who wanted to get into the White House'' (cited in Bradlee 1988, p. 118). But, according to Bradlee (1988, p. 119), General Schweitzer may have ''misjudged the whirlwind energy that Ollie would nonetheless bring to the position.''

At the NSC, North maintained the same frenetic pace that characterized his previous military career: He ''was working an incredible eight-five to ninety hours a week,'' later remarking that ''I love the work, but we have to lift some of this onto the CIA so that I can get more than 2–3 hours of sleep at night'' (Woodward 1987, pp. 292, 468). (The constantly active Mafioso entrepreneur or Ivan Boesky comes to mind.)

According to Roger Fontaine (cited in Bradlee 1988, p. 124), who headed the Latin American section of the NSC when North got there, roles for beginning staff were not ''terribly clearly defined'': ''It's a job for self-starters.'' North volunteered to help Fontaine in the area of Central America. When asked during the Iran-Contra Congressional hearings who gave him the job of managing the administration's assistance to the *contras*, it is revealing that North (1987, p. 105) replied, ''Well, I guess it fell to me by default.'' In addition, North's other assignments had an action-oriented cast to them,'' since they involved crisis management. While in the absence of crisis, such work can be quite routine and uneventful, ''action tended to follow North around,'' for the beginning of 1982 saw ''a dramatic increase in terrorism'' against Americans (Bradlee 1988, pp. 129–33).

By 1984, North was also consumed by the administration's effort to help the *contras* in Nicaragua: ''He virtually abandoned his family, often sleeping a couple of hours on the couch in his Old Executive Building office, and there were many occasions when he did not sleep at all.'' His trips to Central America were an exciting adventure precisely because of his self-selecting action-orientation: ''His methods were the same in every aspect of his life: Get as close to the edge as possible and run the greatest risk, so that the thrill of survival would be as intense as can be'' (Ledeen 1988, p. 78).

Ironically, when in accord with normal rotation, North was advised by the Marine Corps that it was time for him to move on—and upward professionally—to a tour with the Naval War College, North now ''was not interested'': ''He had developed an appetite for being where the action was, the terrorism account was getting busier, and he was consumed by the *Contra* passion'' (Ledeen 1988, p. 86).

Two observations conclude this section: (1) North did not ''develop'' an appetite for action while at the NSC; rather, he had a biological propensity for action, *beginning with conception*. The NSC succeeded in feeding that appetite only because his self-selecting tendencies encouraged that result; and (2) if North was able to thwart the Marine Corps' desire that he leave the NSC—and he was—it was only because of the bonds that he was able to form with other action-oriented men.

Bonds and network formation. If we start with North's tenure at NSC (ignoring earlier bonding such as with Captain Goodwin in Vietnam), the first major bond involved the deputy to the military aide to the national security advisor—Robert ''Bud'' McFarlane (whose superiors then were John Poindexter and William Clark). As Bradlee (1988, p. 26) notes: ''With Annapolis, Vietnam and the Marine Corps in common, it seemed natural that McFarlane would gravitate toward North and vice versa.'' Senior in both rank and age to North (though only six years older), McFarlane developed something of a father-son relationship with him. In McFarlane's own words, he was impressed that ''North was an extreme example of workaholicism.'' But since ''people always gravitate to someone who'll do the work,'' he also impressed others ''as a can-do person.'' North's advantage was in the sense of energy he communicated.'' In addition he, McFarlane, and North shared ''the same kind of values and priorities'' (cited in Bradlee 1988, pp. 127–28). It was McFarlane who succeeded in extending North's term at the NSC in 1984.

North met a ''kindred spirit'' and an important contact in Texas billionaire H. Ross Perot in 1982; in early 1983, North and ''professional adventurer,'' Jack Wheeler, ''hit it off immediately'' as ''a couple of can-do guys who wanted'' to fight the *Contras*; another right-wing idealogue and adventurer Kevin Kattke also received a good deal of North's attention, perhaps because he ''saw a bit of himself in Kattke—a can-do, action-oriented guy with plenty of moxie'' (Bradlee (1988, pp. 130, 154, 180).

Probably the most fateful of North's later friendships was that with CIA director William Casey. As with North, Casey's own career reveals a process of self-selection in part designed to maximize his action-orientation. During World War II, for example, rather than take a desk job, as Casey relates, ''I itched for action and the OSS seemed the best place to get it'' (Casey 1988, p. 4).

Recruited into the OSS by William Donovan, Casey was taken with and influenced by Donovan and his style. As described by Volkman and Buggett (1989,

p. 213), "Donovan was pure action, and sometimes accused of loving action for action's sake." but "what most impressed Casey about Donovan was his ability to get things done." "Casey became close to Donovan and often accompanied him on his assorted whirlwind tours to the OSS outstations (Casey later repeated this habit when he became DCI [Director of Central Intelligence] and was known derisively around some sections of the CIA as 'the wanderer.')"

Casey's own directorship of the CIA was action-oriented. "We are the action agency of the government," Casey reportedly stated (cited in Woodward 1987, p. 417). "As an activist," Menges (1988, p. 368) observes, "Casey was inevitably drawn to individuals (like North) who were also activists." So it was that by 1984 North had established a firm, friendly contact with Casey:

> More and more Lieutenant Colonel North found himself making that short trek from his own office to Casey's. Casey was not a boss but a soulmate. The DCI had evolved into a father figure, an intimate and adviser. He had become a guiding hand, almost a case officer for North. (Woodward 1987, p. 466)

Shadow networks. "Ollie North," writes Smith (1988, p. 617) "was chief executive officer of a covert private sector empire—known among insiders as the 'enterprise'—which short-circuited the U.S. government. As Reagan's case officer for the Iranian operation and the *Contras*, North built a network of former CIA officials." According to Marshall, Scott, and Hunter (1987, p. 15), the "common experiences" of this "elaborate old-boy network of retired CIA and military covert operators who carried out Reagan's policies in the field" ran "the gamut from the CIA-sponsored war against Castro in the early 1960s, to the covert war in Laos later in that decade, to shady arms and intelligence operations in Iran by the mid-1970s." "Out of these experiences came shared expertise, close-knit contacts and trusting friendships that would bring them together again as a covert network in the 1980s." Theodore Shackley, the CIA operative who directed anti-Castro action following the Bay of Pigs, later in the mid-1960s took charge of the "secret war" in Laos; one of his assistants, anti-Castro veteran Felix Rodrigues, later became in charge under North of the *contra* air supply operation in El Salvador; Tom Clines later came close to being implicated with Shackley and Richard Secord in the shady dealings of former CIA operatives Edwin P. Wilson and Frank Terpil, notably illegally selling arms to Libya in 1976.

Woodward (1987, p. 468) states that in 1984 it was Casey who "instructed North to set up a private entity" outside the government in order to supply the *contras* and recommended General Richard Secord to head it. A retired Air Force major general, Secord had also served as a CIA operative in Laos. Secord, after accepting North's offer, soon began bringing in his own assistants, including Tom Clines, whom he had met in Laos, and Raphael "Chi Chi" Quintero, anti-Castro Cuban CIA operative from the Bay of Pigs and a long-time friend of Clines and Edwin Wilson. As Bradlee (1988, p. 198) observes: "Though the tapping of Secord and, by extension, the equally Wilson-tainted Clines and Quintero would years later be assailed as an incomprehensible return to a discredited old-boy network and an irresponsible privatizing of foreign policy, to North and Casey it seemed only logical. They, after all, were the true believers on the frontlines while a spineless Congress shackled them.''

North's ''Enterprise,'' according to Emerson (1988, p. 217) ''was partly a reincarnation of the Pentagon's special operations network. In addition to Secord, other military convert specialists included returned Air Force Colonel Richard Gadd and retired Air Force Lieutenant Colonel Robert Dutton, who handled logistics for the air supply to the *contras*.'' Emerson also observes that the network ''grew at a fantastic rate'': ''Within two years the Enterprise had at its disposal five planes, an air field, a boat, arms warehouses, Swiss bank accounts and numerous companies around the world.''

At the center of the network to aid the *contras*—North called the operation ''Project Democracy''—was the North-Secord bond: ''North liked Secord. The General was a no-nonsense, action-oriented patriot who had disdain for timid bureaucrats and was prepared to go over them to get things done. He and Ollie were cut from the same cloth'' (Bradlee 1988, pp. 198–99).

North made his own additions to Project Democracy. To serve as his personal liaison to the *contras*, he picked from the staff of Senator Dan Quayle a legislative aide, Robert Owen, who was deeply impressed with ''North's zeal'' and who would later testify before Congress that he came to love North ''like a brother'' (Bradlee 1988, pp. 200, 201, 488).

Encouragement from the top. In a book published on the eve of Iran-Contra, Prados (1986, p. 378) notes that the paramilitary action against Nicaragua began with a presidential finding almost immediately following President Reagan's inauguration. In March 1981, Reagan personally approved a covert action against the Sandinistas: ''Not only has the president made the required decisions, he has lent his personal influence to the operators in covert action''—a fervor fully shared and indeed encouraged by his DCI, William Casey. Indeed, Prados (1986, pp. 367, 378), asserting that there ''has rarely been a closer team of President and OCI'' than under Reagan and Casey, concluded that ''what is striking about the Nicaragua campaign is the great fervor with which it has been conducted.''

This presidential encouragement was coupled with ample opportunity. Given Reagan's ''laid-back'' managerial style and inattention to detail, his staff had maximal freedom to implement the general policy that he had laid down (cf. Smith 1988, ch. 16). The loose structure of the NSC, as already noted, further encouraged the initiative of an activist like Oliver North, who, according to Menges (1988, p. 198), was given a freer and freer rein,'' first by McFarlane and then Poindexter, further fueling his ambitions.

The resulting synergy. Smith (1988, pp. 626–33) lists four reasons for the National Security staff's ''stunning ascent'' to power under Reagan, as well as that of Oliver North. First was the president's ''hands-off'' managerial style, which permitted the second reason: ''An assertive, activist national security staff, taking general cues from the president . . . great leeway to speak for the president.''

This second reason included the activities of McFarlane, who initiated the policy of attempting an opening to Iran, and John Poindexter, his successor, who continued it. The third reason for North's rise to power (discussed more fully below) was Bill Casey. The fourth element in North's rise was "the extraordinary leverage he gained by force of personality, by his cocksure manner, by acting when others hesitated." These last two reasons correspond to our emphasis on the self-selection of like-minded activists into critical positions under optimal circumstances.

In examining North's relationship with Bill Casey, Smith observes: "North was in awe of Casey as a spymaster and intellectual. And Casey evidently admired North's can-do daring, and saw him as a soul-mate, perhaps even as a son. They thought alike, and they were in almost constant contact. Records showed that they had at least thirty-five meetings, and probably many more" (Smith 1988, p. 633). The bonding of think-alike activists, as indicated in the case of Hunt-Liddy with Watergate, can have synergized effects. The Casey-North partnership seems also, as suggested by Smith, to illustrate this likelihood.

Gutman (1988, p. 314) also notes that the "Casey-North style suited Reagan's preference for action over deliberation, but their rapid accretion of influence should be seen in the context of the more pessimistic environment in which they operated." For Gutman, a number of personnel changes were critical. In perhaps the most important of these changes, Donald Regan switched positions with James Baker to become the new chief of staff: "A Casey protégé had replaced a Casey rival. The former head of a Wall Street investment firm whom Casey had brought into the administration, Regan had a limited background in foreign policy" (Gutman 1988, p. 314). Therefore, Regan was more likely to go along with initiatives originating from Casey and indirectly from North.

With this as background, we will examine the fateful decision to divert funds illegally from the sale of arms to Iran to aid the *contras*. According to North before the Tower Commission, the idea was first suggested by Amiram Nir, the action-oriented advisor for counter-terrorism to Israeli prime minister Shimon Peres (*Tower Commission* 1987, p. 53). According to North before the Iran-Contra Committee of the U.S. Congress, it was rather suggested to him by Manucher Ghorbanifer, the Iranian entrepreneur who had served as one of the intermediaries between the United States and Israel, on the one hand, and Iran, on the other. Both of these accounts, however, are denied by Israeli sources, which implies that North initiated the idea (see Segev 1988, p. 241).

In any event, according to North's congressional testimony, North first shared the idea of diverting Iranian funds to the *contras* with Casey: Casey loved it, North reported, thinking it "the ultimate covert operation" (cited in Smith 1988, p. 630). North's well-known penchant for exaggeration and prevarication always has to be taken into account. And William Casey did not live to present his own version. Indeed, we do not even know if Casey was aware of the diversion, though Woodward (1987, p. 507) attests that he was. But some facts are indisputable: North not only knew of the diversion, he strongly favored and implemented, if not initiated, it. And he was not alone but was supported by the synergistically charged connections that he had helped form.

We will now consider an even more intriguing question: Did Casey in fact, propose or favor an unsupervised covert operation? Woodward (1987, p. 467) accepted North's contention that Casey favored an ''off-the-shelf,'' independent, ''self-financing,'' and ''full service covert operation.'' North's assertion concerning Casey's proposal has been rejected by at least one Casey defender (cf. Menges 1988, pp. 366–71). Woodward (1987, p. 505), however, concludes that as much as anybody, Casey was ''behind the contra operation, the Iran initiative and the range of other secret undertakings and clandestine relations ''and that he once told Woodward that his ambition was to prove that his country could ''do these things,'' meaning, Woodward believes, ''covert actions conducted in true, permanent secrecy.''

This ''off-the-shelf'' independent covert action capability, like project Gemstone from the Hunt-Liddy collaboration, represented the ultimate synergistic spinoff from the Casey-North bond, at least verbally. Although Casey's role is unsubstantiated and known only through North's admittedly untrustworthy testimony, a number of points are worth noting. First, the illegal operation allegedly advocated by Casey is sufficiently in character, so it is plausible that he did propose it. Of course, it is entirely conceivable that North, not Casey, actually initiated the idea. In any event, North approved of it. At face value, North's testimony is not at all outrageous.

Second, as an activist, Casey was frequently impatient of bureaucratic inertia in established agencies, including his own CIA (cf. Woodward 1987, pp. 382, 388, 417). Third, it was Casey, who had, as already indicated, apparently advised North to set up a private network to be headed by Secord to aid the *contras*.

Fourth, it appears that the Casey-North connection was quite close. Some of Casey's top subordinates noticed ''the director's close relationship with North,'' with their constantly phoning one another (Bradlee 1988, p. 129); once when Secord was to meet with North for dinner, he gave up waiting after ''Casey collared Ollie alone'' for at least an hour (Bradlee 1988, p. 328). Unlike North's unsupported boasts of one-on-one meetings with President Reagan, this relationship was an intimate one—with one of the most powerful figures in the Reagan administration.

One final note: It is important to remember that during the last two years of his life William Casey had a brain tumor. Anderson (1988, p. 348) emphasizes that this period, culminating in Casey's death on May 7, 1987, following a brain operation several months before, corresponds precisely with the critical unfolding of Iran-Contra.

Anderson (1988, p. 349) notes the possibility that ''the onset of a brain tumor can cause drastic changes in a person's judgment and personality.'' The type of brain tumor Casey had was ''particularly insidious,'' according to Dr. William R. Shapiro, professor of neurosurgery at the Memorial Sloan-Kettering Cancer Center in New York ''because it grows like a network within brain tissue rather than in a ball or lump. . . . As a result, the tumor can induce subtle changes in personality or mental functioning that may go unnoticed by associates or may be attributed to stress until there is a dramatic symptom such as a seizure'' (cited in Anderson 1988, pp. 349–50). Such a seizure befell Casey in December 1986. Personality changes were detected by a close friend of twenty years, columnist William Safire: ''He grew testy with friends, easily infuriated at criticism and hypersensitive to leaks'' (cited in Anderson 1988, p. 350). Anderson (1988, p. 351) surmises: ''There is a good chance that by the time the Iran-Contra affair was under way the wise old bull of the intelligence community was transformed by an invisible assassin into the equivalent of a crazed, rogue elephant, driven wild by an unseen tormentor, rampaging and destroying whatever lay in its path.'' (Menges does not consider this possibility at all in defending Casey from North's testimony.) Anderson, however, may have overstated the extent of the ''transformation.'' First, Casey's career had been consistently characterized by a notable action-orientation; second, the gradual, subtle changes in personality or mental functioning that may occur are not necessarily sharp or dramatic. What we can surmise is that a key actor in the Iran-Contra affair had his natural action-orientation intensified and distorted from within in the form of a cancerous brain tumor and from without in the form of close connections with an equally activist Oliver North. There is an underlying biological dimension here, then, genetically—the influence of self-selecting tendencies—and physiologically—the impact of Casey's brain tumor.

IMPLICATIONS

In considering the careers of E. Howard Hunt and Oliver North and their involvement in two of the most well-known government scandals in recent American history, some long-term implications need to be considered if the self-selection process outlined in this text has any validity. First, in the words of Prados (1986, p. 363; emphasis in the original), ''What do you do with a *former* secret warrior?'' This problem may be accentuated by special circumstances. For example, although Allen Dulles cited the ''disposal'' problem to John F. Kennedy ''as one reason to go ahead with the Bay of Pigs invasion,'' far ''from disposing of the Cuban contract agents, the operation led to almost open-ended CIA involvement with the Cuban exiles'' (Prados 1986, p. 363); the shadow networks formed spanned the Bay of Pigs through Watergate, Iran-Contra and potentially beyond.

Controlling or reducing covert action, however, may only serve to drive it under ground. As Marshall, Scott, and Hunter (1987, p. 20) note, the reforms under the Carter administration to cut back on covert action may have ''contributed to a more pervasive Iran-Contra scandal,'' in part by leading to extra-governmental shadow networks made up of covert operators either forced out of or disillusioned with those agencies responsible for covert action. Thus, former CIA operative Edwin P. Wilson's conviction for conspiring to trade in arms with Libya and for conspiring to murder two U.S. prosecutors and six other people resulted in several dismissals and resignations from the CIA, but this later result indirectly contributed to the formation of the shadow network that was to center about the business dealings of Richard Secord, the retired air force general who was to work for Oliver North.

As a result of internal reforms within the U.S. government, a dramatic rise in private sector activity in covert operations has occurred. The extra-governmental role of Richard Secord in Iran-Contra is simply the most dramatic illustration of this trend (see Brown and Arnold 1988). The revolving door between the public and the private sectors increasingly includes the realm of cover operations. It is in these interstices that the shadow networks form and perpetuate themselves. These extra-governmental networks can support governmental covert operations, as in Iran-Contra, or as Houghan (1978) has detailed, they may serve such private clients in the manner that former FBI agent Robert Maheu and associates served Howard Hughes. "Private CIA's-for-hire," writes Houghan (1978, p. 15), have, following World War II, "metastasized across the landscape."

Moreover, there may be what Prouty (1973, p. 280) refers to as "additional thousands of CIA camp followers": These include members of the academic, publications, and business world "who enroll themselves or who have become enrolled for various reasons in the lure of 'fun and games.'"

It is well to keep in mind the underlying assumptions made here to account for these developments; for one can agree with their description but assign quite different causes.

First, a process of self-selection operates in which the genetically influenced trait of action-orientation plays a critical role. An opposing view is an environmentalist perspective. In accounting for the Bay of Pigs and Watergate, for example, that is precisely what Janis (1982) takes with his concept of "groupthink," the notion that members of cohesive groups are led to conform to the norms, values, and attitudes of the group. But this notion, good so far as it goes, ignores the possibility that the cohesiveness of the group might be due to the compatibility of individuals sharing similar genetically influenced traits.

One illustration may be added in reference to Iran-Contra, or more precisely, in reference to the secret mining of Nicaraguan harbors in 1983. According to Gutman (1988, p. 146), the "process of formulating and implementing this policy" bore the earmarks of "groupthink," alluding explicitly to Irving Janis. Indeed, Gutman (1988, p. 194) calls attention to the mining as "an example of the perils of formulating policy in a small, secretive, and homogeneous group," referring to the Restricted Interagency Group (RIG), made up of representatives of the State Department (Langhorne Mosley), the CIA (Dewey Claridge), the National Security Agency (Oliver North) and the Joint Chiefs of Staff (Arthur Moreau, Jr.). Gutman goes on to observe that the "foursome who comprised the innermost policy circle—Motley, Claridge, Moreau, and North—were action-oriented men with military or paramilitary backgrounds who had no compunctions about covert operations." The president's representative from the NSC, Gutman feels, should have provided a critical review, with "genuine alternative actions to the President for decision," but the NSC representative "was not a disinterested party" for "North was a zealous advocate" (p. 198).

In evaluating North's tenure on the NSC, Gutman (1988, p. 207) observes: "Three years at the center of the policy process had neither broadened Oliver North's vision nor tempered his method of operation. His whole persona remained that of the action-oriented lower-echelon infantry officer who would batter down walls to carry out his mission." In other words, North's environmental milieu neither broadened nor moderated his life style. Here, then is the critical qualification of "groupthink": In his covert involvements, North did not just passively conform to the overall sentiments of his colleagues; rather, as a "zealous advocate" (Gutman 1988, p. 198), he intensified the view of his like-minded colleagues.

Second, the formation of bonds and networks develops through the self-selection of like-minded individuals. In his study of the burgeoning private realm of covert operations, Houghan (1984, p. 22) states, "Many of the individuals and firms discussed have relationships with one another—it's an incestuous universe." As Prados (1986, p. 364) puts it, "Retired intelligence and military types have maintained many international connections and work in areas similar to what they did in the service."

Third, action-oriented individuals, particularly those engaged in covert activities, are likely to be less morally inhibited. That may be one of the subliminal appeals of covert action. As Oliver North testified before the Iran-Contra committee (and the American people): "By their very nature, covert operations . . . are a lie." Concerning the recruitment of Ghorbanifer to serve as an intermediary with Iran, North testified: "I knew him to be a liar, I knew him to be a cheat . . . we knew what the man was, but it was difficult to get other people involved in these kinds of activities. I mean, one can't go to Mother Theresa and ask her to go to Tehran" (cited in Bradlee 1988, p. 332, 499).

This morality dilemma is described by Colby (1978, pp. 86–87). As a young covert operator, well before becoming CIA director, Colby observed:

> But perhaps the most complex lesson I had to learn—and one that really could not be taught in training or anywhere else—was how to lead the double life. And I have to say that there was an enormous temptation to not even try. Considering the importance and all-consuming nature of the work I was doing at the Agency; considering the missionary zeal, sense of elitism and marvelous camaraderie among my colleagues there; considering above all that I was strictly forbidden to talk about what I was doing with anyone outside the Agency and thus couldn't share my concerns or just sit around shooting the breeze in shop talk with anyone in the world—considering all of this, one can see how easy it would have been for me to drop out of that world and immerse myself exclusively in the cloak-and-dagger life. And some of my colleagues at the Agency did just that. Socially as well as professionally they cliqued together, forming a sealed fraternity. . . . In this way they increasingly separated themselves from the ordinary world and developed a rather skewed view of that world. Their own dedicated double life became the proper norm, and they looked down on the life of the rest of the citizens. And out of this grew what was later named—and condemned—as the "cult" of intelligence, an in-bred, distorted, elitist view

of intelligence that held it to be above the normal processes of society, with its own rationale and justifications, beyond the constraints of the Constitution, which applied to everything and everyone else.

Given the framework presented here, we can assume that a process of self-selection is responsible, in large part, for the presence of so many colleagues with a missionary zeal for work of an all-consuming nature although, perhaps, "mutual selection" might be a more apt expression, given the rigors of the CIA process of personnel selection. These personnel, sharing similar personality traits and political outlooks, find themselves compatible socially as well as professionally. They, therefore, especially given the other incentives Colby alludes to, bond to form "a sealed fraternity." The resulting social environment intensifies the missionary zeal of its members. The end product is the "cult of intelligence" mentality, with all of its amorality and political dangers.

In making these underlying assumptions, we must keep in mind a populational framework. Any large population or organization will contain a diversity of people. Thus, certain governmental agencies—the CIA, the FBI, the marines—may well disproportionately attract action-oriented individuals; on the other hand, these organizations are sufficiently large to include slots that neither require nor attract activists. We might note that William Casey more than once felt frustrated by subordinates within his own CIA.

A biologically grounded approach, moreover, does not exclude the influence of environmental factors. If Oliver North had been killed instead of wounded in Vietnam, Iran-Contra might have taken a different turn. But given an Oliver North, a biologically grounded self-selection process may be critical to understanding Iran-Contra: An activist like North might be assigned an unwanted desk position at the National Security Council and proceed to invest it with all of his considerable energy and to ally himself effectively with others of like bent.

It is well to reiterate that action-oriented individuals will self-select in different ways, depending on both social circumstances and other genetically linked aptitudes. Future research will have to explore the parameters of these relationships.

One suggestive comparison from Japanese history comes from Craig (1970, p. 1):

> There appear to be striking similarities among the *Shishi*, the proemperor fanatics of the late Tokugawa era, certain groups of ultra nationalist young officers in the 1930s, and the anarchic leftist student groups during the late 1960s—in spite of their different ideologies and differing historical contexts. In each case radical ideas and feelings of moral righteousness "cooked" within the group, ego reinforced alter, pressure built up, until action erupted in the form of assassinations or group violence.

It is not hard to see the operation of self-selection and hypertrophy at work here. Even genetic factors are implied, for Craig also asserts here that, "Japanese energy or drive cannot be explained solely in terms of social or cultural factors."

If action-orientedness is a biologically based propensity, it has presumably existed throughout human history. Accordingly, the self-selection process outlined here must have operated in times past. Indeed, Marshall, Scott, and Hunter (1987, p. 26) suggest that a "disposal problem" has existed throughout history: "In the fifteenth century a problem much like this one—what to do with ravaging knights who lived off the land—was a powerful motive for England to prolong the Hundred Years' War and keep its knights in France."

Let us examine the possible political ramifications of action-orientation within the broadest framework. Schumpeter (1951) analyzed expansion by peoples undertaken for its own sake, that is, imperialism; *pace* Marxist analysis, he did not view imperialism as an extension of economic class interests. Rather than being influenced by "real" or "concrete" interest, Schumpeter (1951, p. 83) saw imperialist tendencies as existing throughout human history stemming from "non-rational and irrational, purely instinctual inclinations toward war and conquest." Schumpeter (1951, pp. 83–84) proceeds to replace this biological explanation (instinct) by a social-Lamarckian one: "The explanation lies, instead, in the vital needs of situations that molded peoples and classes into warriors—if they wanted to avoid extinction—and in the fact that psychological dispositions and social structures acquired in the dim past . . . tend to maintain themselves and to continue . . . long after they have lost their meaning and their life preserving function." Thus, from this perspective the force behind an imperialist policy is an atavistic one, with those directly involved in its pursuit as warriors coming to perceive an interest in maintaining such a policy, an interest that is largely non-economic.

Although the thrust of Schumpeter's foundation is quite compatible with the basic analysis of this paper, its social Lamarckian etiology is not. According to MacDonald (1988a, ch. 1), four different models of "nature-nurture interrelationships" can be distinguished: (1) an environmental determinism; (2) a weak interaction relationship in which basically passive organisms are differentially affected by the same environments; (3) a strong interaction model, in which a dynamic reciprocal relationship develops between the organism and the environment, the organism initiating action and affecting the environment as well as vice versa; and (4) a genetic determinism. Although MacDonald properly points out that no one model can possibly accommodate all of the observed behaviors of individuals—each individual and each trait conceivably demanding different treatment, neither a genetic *nor* environmental determinism is likely to suffice in accounting for most behaviors. As Masters (1989, preface) emphasizes, such an "either or" perspective is not generally appropriate for the study of human affairs.

An emphasis on self-selection would fit the third or strong interaction model, keeping in mind that a populational perspective would qualify the application. If, for example, an action-oriented continuum—from the very passive to the very active—were applied to any large population, the most passive individuals might better fit the weak interaction approach, while the highly action-oriented would clearly fit the strong interaction position.

Keeping in mind that the available evidence suggests that action-orientedness is genetically influenced and that it is significantly associated with aggression, that is, more action-oriented individuals tend to be more aggressive in behavior (see MacDonald 1988a, pp. 195–97), I will now briefly reformulate Watergate and Iran-Contra to correspond with Schumpeter's general theory of imperialism.

First, the perceived threat emanating from the newly revolutionary government in Castro Cuba leads an activist DCI, Allen Dulles, to take action that will mobilize Cuban exiles, a covert operation that will culminate in the abortive Bay of Pigs.

Postulating a genetic dimension underlying this operation does not exclude the possibility that real political and economic motivations also played critical roles, including patriotism and material self-interest. But note that those most likely to be actively attracted to such a mission would share certain biologically based characteristics—for example, younger males as opposed to women, children, and the very old.

> In the mafia area, acts, people and events were habitually judged in terms of the honour acquired through victory in struggles and competitions. It follows that aggressiveness and violence were positively endorsed. In consequence, those who for biological and cultural reasons were least disposed to take part in conflict (such as women, children, and the elderly) found that their participation in the life of the community was severely limited—unless they adopted the manly and pugnacious bearing of those in the superior categories. (Arlacchi 1986, pp. 12–13)

(Arlacchi proceeds to observe that in his description of one of the female activists in the arena of mafia competition her son emphasized "her strength, her courage, and her rather bloodthirsty temperament." It is noteworthy that no women seem to have occupied important roles in Watergate or Iran-Contra, although Hunt's wife, Dorothy, had been a CIA agent and her death in a plane crash might have been related to the scandal).

Second, the buildup of covert operation networks, once formed, tend to perpetuate themselves. Again, ideological and material reasons may play a part here. Also the experience of having engaged in a common enterprise may prove reinforcing. But all of these explanations, by themselves, are inadequate. As we've seen, given a common ideological motivation, not all individuals in a given population will respond equally. And although those who do respond may then acquire a personal stake in the activity, that likelihood does not explain why, when individuals leave an activity, their self-selecting propensities persist. In each case, they look for and find the kind of activities that they find congenial. In the process, covert activity may become extra-governmental and even illegal; but it persists. Previously established networks do not disintegrate but are kept alive by personal contacts and new ventures (such as in Edwin P. Wilson's or Richard Secord's business dealings). They become shadow networks that can be reactivated for other purposes. Both Watergate and Iran-Contra illustrate such an eventuality.

This last point bears emphasis. A quiescent period in governmental covert operations—such as between the Nixon and Reagan years—does not mean that the potential for such action had disappeared. For purposes only of analogy, take Ryan's (1984) study of "Quiet Neighbors"—that of former Nazi war criminals who escaped to the United States and with scarcely an exception seemed to become "model citizens." Obviously changed historical circumstances had a decisive impact on the outward behavior; but it would be naive to imagine that they had now been totally remade by a new environment. (Furthermore, it would be of interest to learn if some self-selecting tendencies persisted under a respectable cover; for example, if any became avid sportsmen, such as zealous hunters.) The larger point, in any case, is that as important as historically changed circumstances may prove, and they may indeed override (or more likely obscure) genetically based tendencies, they do not simply eliminate them. It is possible that these tendencies might materialize (or reappear) in some collective form to effect history (for example, Watergate and Iran-Contra), through the process of individual self-selection that comes to involve numerous like-minded individuals.

We return, then, to E. O. Wilson's concept of hypertrophy, the process wherein the social environment reinforces pre-existing propensities. As Barash (1986, ch. 7) has noted, this has led, in the realm of aggression, to the possibility of the end of man through nuclear destruction. Obviously hypertrophy, if it led to such an extreme eventuality, would indicate that not all biologically based traits, *pace* a sociobiological premise, may prove to be adaptive. Such may especially prove to be the case with human society, evolving from small homogeneous isolated populations to a densely populated world that is both interconnected and heterogeneous.

Thus far, at least, the terror of a nuclear war has proved to be its own effective deterrent. But since 1945, a number of conventional wars have occurred, for example, the Korean, Iran-Iraqi, and several Israeli-Arab conflicts in which a superior technology has indeed often been a decisive factor. And, of course, frequent guerrilla wars have broken out—notably the protracted one in Vietnam from 1946 through 1978, first against the French and then against the Americans—in which a superior technology may not prove decisive.

Paramilitary covert operations, including terrorism, have also been frequent occurrences. These activities, too, make use of all the latest technology that is appropriate and available. The Liddy-Hunt Gemstone plan, for example, envisioned the use of electronic surveillance, including a "chase plane" to intercept radio-telephone communications (Lukas 1976, p. 172). Iran-Contra involved shipping arms to Iran and using the profits (or rather some of them) to aid the *contras* with arms. Terrorists, of course, employ the latest technology, including the paradox of "sophisticated low-intensity terrorism" that is, antitechnology terrorism in which they employ the most advanced technology against such advanced technologies as biotechnical institutes or chemical warfare plants (see Mullen 1988 and Wiegele 1990). Thus the appropriate technology for terrorism or paramilitary covert action appears steadily to advance, while the persistent urge to use it, as Schumpeter suggested, remains atavistic.

Part II

The Neuropolitics of Local-Cosmopolitan Self-Selection

Chapter 4

Locals, Cosmopolitans, and Politics: A Neuropolitical Perspective

> When *Socrates* was demanded whence he was, he answered not "of *Athens*," but "of the world"; for he, who had his imagination more full and further stretching, embraced all the world for his native City, and extended his acquaintance, his society, and affections to all mankind, and not as we do, that look no further than our feet.
>
> —Montaigne, *Essays*

> It is one thing to see the forward path and another to be able to take it. But it is better to have an ambitious plan than none at all.
>
> —Winston Churchill, *The Second World War*

HUMAN TIME AND TERRITORIALITY: THE SPACE-TIME HORIZON

The wider a person's space-time horizon, the greater control over his or her life and the lives of others. This is the central theme explored in this chapter. Space-time horizon is defined here as the capacity to transcend conceptually one's immediate place in space and time. Referring to time, this capacity involves "a person's ability to act in the present in the light of his anticipation of relatively distant events." Cottle and Klineberg (1974, p. 16) see this ability as entailing "three central processes that are uniquely developed in human beings": "(1) the prodigious capacity to manipulate symbolic representations of reality, and thereby create images of absent events and believe in their validity; (2) the ability to interpret the symbolic representations into ongoing patterns of actions and meaning, . . . and (3) the present experiencing of pleasure or discomfort generated solely by these symbolic representations.''

As for space, the capacities involved in a person's horizon are those which Gardner (1983, p. 173) views as central to spatial intelligence: ''the capacities to perceive the visual world accurately, to perform transformations and modifications upon one's spatial perceptions, and to be able to re-create aspects of one's visual experience, even in the absence of relevant physical stimuli.'' Gardner proceeds to note these various abilities are ''clearly not identical''; for our purposes, it is spatial intelligence in relation to the ability to represent the visual world accurately that is most important.

Malmberg (1980, p. 5) observes that human spatial behavior is ''strongly dependent on the individuals' cognitive map of the environment.'' The spatial behavior and cognitive map relate to an individual ''orbit'' or ''human range,'' that is, the total spatial area through which a person habitually or occasionally roams and therefore is in some sense occupied (see Malmberg 1980, p. 2). Note that these terms correspond to a broad definition of territoriality, but *not* to a narrower meaning involving, for example, the personal space or specific estate of the individual. Unlike more private forms of territoriality, an orbit or home range is necessarily a public form, shared by numerous people (unless one is speaking of a hermit or recluse).

Personal and Political Mastery

As stated at the outset, the broader a person's space-time horizon, the greater his potential control over his own life and the lives of others. As Weissman and Gorman (1977, p. 41) put it, ''Insightful, directed activity requires conceiving of an extended spatiotemporal field, proceeding from given positions toward intended destinations.'' Or as Cottle and Klineberg (1974, p. 73) assert, ''It is man's prodigious capacity to create and manipulate symbolic representations of physical and social reality that underlies the uniquely human proclivity for anticipating distant future events and acting on them.''

The clearest way to illustrate our central theme is by imagining an individual who for some reason cannot conceptually transcend immediate time and space—a two-year-old, for example, or a severely retarded or brain-damaged person. Such an individual is simply unable to conceive of alternative courses of action through space and time; therefore he or she is especially prone to become a creature of the immediate environment (in space-time). It is not incidental that Skinnerian behavior modification is most effective on younger populations and those least capable of cognitive representations of their world.

A relatively broad space-time horizon may therefore serve as a necessary condition for effective mastery over one's life and those of others. But it does not, however, serve as sufficient condition. A possession of a capacity does not necessitate its fulfillment.

FOUNDATIONS

Evolutionary

The uniquely abstract space-time horizon of the human species appears clearly to be the result of an evolutionary process. "In evolutionary perspective," Weissman and Gorman (1977, p. 4) write, "more active and differentiated goal-seeking and danger-avoidance appears linked to the greater adaptive 'intelligence,' with greater awareness of the surrounding world and increased capacity for appropriate response."

Gardner (1983, p. 184) speculates on the evolutionary basis of territorial awareness: "On nearly all accounts, spatial intelligence assumed pivotal importance for a roving band, whether it was involved in gathering or in hunting. When individuals needed to traverse wide spaces and return safely to their homes, it was important to have a keen spatial intellect; otherwise, the chance of getting lost would be far too great."

Developmental

Piaget (1955, 1966) has shown the developmental basis for an abstract space-time horizon. Not until early adolescence, when a child's intellectual development proceeds from the stage of concrete operations to that of formal operations, can an individual become conceptually free of the immediate or concrete world. As Flavell (1963, p. 223) summarizes: "The child deals largely with the present, with the here and now; the adolescent extends his conceptual range to the hypothetical, the future, and the spatially remote."

With respect to time, Cottle and Klineberg (1974, p. 86) state: "The acquisition of formal operations makes it possible for an adolescent to extend the span of future time that he is able to integrate with his present experience." As a result, he is able "to deal realistically and logically with conceptions of the distant future, with images and possibilities whose only reality is in the words he says to himself" (Cottle and Klineberg 1974, p. 85).

The conception of space becomes similarly broadened. "Only during the formal operation era, at the time of adolescence," Gardner (1983, p. 179) writes, "can the youth deal with the idea of abstract spaces or with formal rules governing space." Only with adolescence, Gardner (1983, p. 180) further points out, is the youth typically able to draw an explicit map of territory that he or she otherwise "knows" on a practical everyday basis.

Neurobiological

A bird brain and a human brain vary in ways both quantitative and qualitative, with immense implications for the perception of the world. The same statement seems to apply for the two-year-old and twenty-two year old brain of the same person.

In any event, the neurobiological basis of the space-time horizon is clearly suggested by prefrontal lobotomies and by other instances of brain damage. The frontal lobes, Cottle and Klineberg (1974, p. 15) write, are implicated in the ability "to maintain response set over a ninety-second delay in the case of the chimpanzee, and to interpose a consideration of long-range past and future events in the case of the human being." Gorman and Weissman (1977, p. 252) also point to the evidence linking damage to the nervous system to "temporal distortions," although they conclude that "no specific brain region has been shown to be uniquely responsible for temporal disorientation." (See also Kohen-Raz [1977, p. 37], Ornstein [1969, pp. 48–52], and Orme [1969, pp. 162–65]).

As for space perception, Benton (1982, p. 270) points out that disorders in spatial thinking associated with disease of the nervous system were recognized as early as the 1880s, and more recently the "paramount importance of the right hemisphere in mediating spatial performance" has been recognized (see also Gardner 1983, pp. 185–90).

THE SPACE-TIME HORIZON FROM AN INTERACTIONAL-POPULATIONAL-MENTALIST PERSPECTIVE

Interactional

If heredity and environment influence nearly every trait, one's space-time horizon is not a likely exception. McGee (1982, p. 203), citing numerous studies, concludes that "accumulated evidence shows that spatial abilities are as heritable as or more heritable than verbal abilities" and "much less correlated with measures of environmental quality such as level of education and socio-economic status."

No such comparable work and conclusions seem to have been produced in the area of temporal capacity. It is hard to imagine that dissimilar results would emerge, for presumably the neurobiological-developmental basis for temporal capacity has a firm genetic foundation. Consistent with the interactional framework here presented, such a genetic basis is assumed.

It is possible, of course, to make a different assumption. Cottle and Klineberg (1974, p. 196) state that "perspectives on the personal future are profoundly shaped by the social and cultural context in which a person is embedded" and that "it is therefore to these cultural and social realities that one must look first for explanations of individual differences in outlooks on the future." We must keep in mind, however, that this conclusion immediately follows a reference to Banfield's controversial characterization (1968) of "lower class culture" as present-oriented.

Two points should be mentioned. First, the cognitive capacity to envision a future and the motivational inclination to work for it should not be confused. Although the motivation may depend on the capacity, the latter may surely exist without the former. It is, in any case, likely that motivational influences are more directly culturally and experientially based than are cognitive capabilities. Second, any discussion of differences in time perspectives, whether cognitive or motivational, should be placed within a populational context.

Populational

Banfield's application of differential time perspectives to different social classes has been critiqued elsewhere (White 1972). The conclusion of the critique is that Banfield (1968, 1974) errs by being too environmentalist and typological. On the one hand, Banfield holds that a person's time perspective is wholly the product of environment and, on the other, that a particular social class background is uniformly the same for all of its members, thereby typologically producing the same time perspective in all. If Banfield were to acknowledge the influence of genetic factors, as well as the profound reality of human genetic diversity, he would have recognized the likelihood that individuals growing up in similar environments may nonetheless respond in different ways.

The following example from the earlier critique (White 1972) illustrates this point: Frederick Douglass as an American slave was expected to have no thought for the future or any plans to escape; but by the time he was a boy of twelve, he was entertaining dreams of his future freedom and escape. As related in his *Narrative*, Douglass (1963, pp. 101–02) found that "in spite of him [his master], and even in spite of oneself, I continued to think, and to think about the injustice of my enslavement, and the means for escape." To escape successfully, as Douglass did, required exceptional space horizon and spatial skills. It is only reasonable to assume that not all slaves, just as not all members of any large population, would equally possess such capability. As Granit (1977, p. 82) postulates, "Normal levels of consciousness in a population" are "determined by a Guassian distribution curve": "Some people seem to maintain a much higher level of conscious awareness than others." What this postulate means, with respect to the space-time horizon, is that a wide range of individual differences is likely to exist with respect to the capability involved in forming the horizon.

Mentalist

If consciousness is an emergent property of the higher nervous system and has causal potency, then, as Popper (1979, p. 229; emphasis in the original) asserts, "What we want is to understand how such non-physical things as *purposes, deliberations, plans, decisions, theories, intentions,* and *values*, can play a part in bringing about physical changes in the physical world." Etkin (1981, p. 67) puts it as follows: "Human behavior is not a completely mechanistic system but is, in part, determined by what is interpreted as part of the future. Our behavior is thus to some extent goal directed, capable of changing direction to keep in line toward something that does not yet exist."

From a mentalist viewpoint, there is an awareness that objective reality and the subjective characterization of that reality are not the same. Each species, as von Uexüll postulated some time ago, has a "world" and an "experience" of its own, and its own "umwelt." "Humans do not exist simply in a physical world," writes Weissman and Gorman (1977, p. 6), "they experience and inhabit a symbolically transformed universe." Human time, for them, becomes "an emergent set of cognitive construct systems based on personal experience and identity, influenced greatly by sociocultural and historical awareness."

Similar observations apply to territory. According to Malmberg (1980, p. 5), "It should be stressed that physical space is . . . experienced by every animal species in a specific and, at least among humans, in an individual manner, too.

The last observation reminds us of the need to keep in mind a populational perspective. Each individual will have his or her own "umwelt"—his or her own awareness and experience of space and time. Consequently, a wide range of individual differences are likely to exist with respect to space-time horizons. With this likelihood in mind, we turn to a consideration of socio-political variables.

THE LOCAL-COSMOPOLITAN CONTINUUM

Individuals vary widely in their mental capacity to represent the world about them. It is not the specificity of detail, in aesthetic or other terms, that is at issue here, but rather the globality of the mental territorial construct.

The two polar types—local and cosmopolitan—that Merton (1957, p. 393; emphasis in the original) introduces into his study of patterns of influence in Rovere (Plainfield, New Jersey) illustrates what is of concern:

> The chief criterion for distinguishing the two is found in their *orientation* toward Rovere. The localite largely confines his interests to his community. Rovere is essentially his world. Devoting little thought or energy to the Great Society, he is preoccupied with local problems, to the virtual exclusion of the national and international scene. He is, strictly speaking, parochial.
>
> Contrariwise with the cosmopolitan type. He has some interest in Rovere. . . . But he is also oriented significantly to the world outside Rovere, and regards himself as an integral part of that world. He resides in Rovere but lives in the Great Society. If the local type is parochial, the cosmopolitan is ecumenical.

Merton (1957, pp. 107–09) further illustrates the differences between locals and cosmopolitans by pointing out their media preferences. Cosmopolitans prefer magazines and newspapers that "devote much of their content to foreign and national affairs," including journals of opinion, such as *Atlantic* and *Harper's*, and the national press, such as the *New York Times*. Locals, by contrast, read the local newspapers and their news preferences within the national news media reflect a desire for a personalized approach. Locals "seek out the personal ingredients in the impersonal array of world news."

Merton's typology must be qualified in at least three ways for our purposes. First, Merton is concerned with describing influentials, whereas we are not so restricted in our analysis. Second, Merton employs two polar types, although he acknowledges (1957, p. 393, fn. 8) that "an intermediate type" might also evolve on further investigation. Given our populational assumption of wide genetic (and environmental) differences in any large population, we have assumed an extensive local-cosmopolitan continuum, ranging from an international and regional cosmopolitan through community and neighborhood locals. This continuum will be discussed in more detail later in the chapter.

Third, Merton does not consider the time dimension in his typology. But it seems plausible to suppose that locals and cosmopolitans differ in their temporal as well as territorial perspective.

At this point, then, we reintroduce Banfield's time horizon. Banfield (1974, ch. 3) distinguishes between present- and future-oriented individuals. Again, we assume for our purposes a wide-ranging continuum with respect to the capacity to orient oneself toward the future, one that is roughly coextensive with the local cosmopolitan continuum just mentioned. Cosmopolitans, in other words, should be much more future oriented than locals. Thus, whenever the local and cosmopolitan distinction is employed, the likely distinctions in time horizon should be kept in mind.

One caveat must be stressed. We are discussing underlying capacities in these space-time continua, not necessarily accomplished realities. Thus, Banfield's typological identification of his time horizon with social class position is not assumed here. In our analysis, it remains an open question as to what the space-time horizon of an individual is, whatever his social class might be. It is entirely conceivable that a child born to royalty or riches will have the capacity for only a distinctly narrow space-time horizon; while some born in the poorest or most oppressive conditions will nonetheless possess the largest capacity. The example of Frederick Douglass illustrates this point. As Quarles (1968, p. 102) writes of the post-slavery life of Douglass: "Douglass liked the company of prominent personages. Himself a travelling lecturer who received the hospitality of others, he in turn entertained anti-slavery speakers who came to Rochester. Celebrities visiting the city were sure to call at the Douglass home." Douglass, in short, was every bit a cosmopolitan, in both orientation and style.

One key variable in whether individuals realize their specific capacities with respect to the space-time horizon revolves around their freedom to bond with like-minded individuals. This variable is discussed in the following section.

LOCAL-COSMOPOLITAN SOCIAL NETWORKS

The primary basis for the social interaction of a local is the proximity of social ties. The primary basis for a cosmopolitan, by contrast, is a highly developed set of interests, which is shaped by acquaintances and which may lead to the formation of distant social ties.

The significance of this distinction between proximity and mutual interest as a basis for social interaction may be seen in the following assertion by Simmel

(1955, p. 30): ''Any association, which is based on local relationships or is otherwise brought about without the individual's participation, differs with affiliations which are freely chosen, because as a rule the latter make it possible for the individual to make his beliefs and desires felt.''

When placing the mutual-interest/proximity distinction within the neuropolitical perspective on bonding that was developed in chapter 2, the key assumption is that individuals tend to bond with others with whom they share relatively similar genetic and neurobiological characteristics culminating in similar perceptions of the world.

The burden of the argument is that ''qualitative,'' as well as ''quantitative,'' genetic similarities contribute to the formation of social ties. From a sociobiological perspective, quantitative genetic considerations are paramount; that is, the greater percentage of genes shared by individuals, the greater their predisposition to cohere. Hence, there is a simple linear progression in this respect from immediate kin through distant kin, similar ethnic group membership to multi-ethnic populations.

An emphasis on qualitative genetic considerations does not deny this progression, but it qualifies it. Genes that influence the design of the brain and so influence our intellect and temperament, as against other genes, are critical in influencing the formation of social ties. Even within the immediate family, in which the numerical sharing of genes is maximal, some ties may be closer than others as a result of unusual qualitative genetically and neurobiologically based similarities. Such qualitative similarities may transcend ethnic lines and national boundaries in promoting social bonds; thus, we may speak of international communities in the arts and sciences in which highly specialized and sophisticated mutual interests are the primary cause. Such communities are highly cosmopolitan in nature.

Similarities in the capacity conceptually to transcend space and time may be one factor in the formation of social ties. Individuals with similar space-time horizons should, given other congenial characteristics, tend to bond; therefore, cosmopolitans will tend to select other cosmopolitans as companions and locals other locals.

''At one extreme,'' Malmberg (1980, p. 223) writes, ''the professional elite is not a primary group based on kinship, ethnicity, nationality or place. Rather it is a voluntary association of men joined by shared interests and values.'' Malmberg proceeds to cite Webber (1964, p. 60; emphasis in the original): ''As men who trade in information and ideas they must necessarily maintain intensive communication with other members of their spatially dispersed, *non-place* communities . . . these folk approximate the true cosmopolites for whom territorial distance is a minor barrier to interaction and whose professional social communities are the last shaped by territorialism.''

At the other extreme, Malmberg (1980, p. 229) asserts, ''social organization'' is ''territorially equal to neighborhood place.'' ''Just outside the few blocks that surround the resident's apartment lies foreign territory. Even though it too may be inhabited by families having apparently identical demographic and cultural characteristics, there is typically no way into a close-knit network of family ties and neighbor associations.'' (Webber 1964, pp. 62–63).

Mutual interest, then, is necessarily a component of cosmopolitan ties and proximity a component of localistic. Nevertheless, the proximity in local relationships suggests a mutuality of interest. At least this is so if a sociobiological perspective is adopted. In this case, people living in close proximity are likely to share a high proportion of genes, thereby providing some basis for mutual interests and outlooks. Furthermore, to the extent that a society is open, cosmopolitans may opt out of close-knit provincial communities, leaving behind those in the group who share a more parochial outlook and who are restricted both in their spatial and social mobility as a result of a limited space-time horizon.

In his study of a close-knit Italian community in the West End of Boston, Gans (1965) appears to describe such a circumstance. Gans (1965, pp. 87–89, fn. 12), directly influenced by Merton's distinction between locals and cosmopolitans, makes a parallel distinction between "person-oriented" and "object-oriented" people. The latter strives for some "object" that "can be conceptualized in a self-conscious and deliberate manner" and whose achievement may necessitate the detachment from some groups or the attachment to others. Thus, "if he is seeking a career he may have to give up childhood friends who are not pursuing one, and join with other individuals who share with him only this common object" (Gans, 1965, p. 90).

Person-oriented individuals, on the other hand, find their goals derived by and limited to the primary groups of which they are a part. If person-oriented individuals have narrow space-time horizons, then it is understandable that they have difficulty in transcending their immediate social environment—their peer groups.

The West Enders, according to Gans, are overwhelmingly person-oriented. Nonetheless, Gans (1965, pp. 254–55) describes the circumstances in which an individual may leave such a "peer-group society."

> Every population—even a low-income one—produces a number of young people with special gifts or talents, be they intellectual or artistic. Among the West Enders, however, young people with such talents are generally ostracized by their peers. And . . . these young people are often forced to choose between their talents and their peers.

"Should he be especially talented and strongly motivated, however," Gans concludes, "the peer group ostracism and the incentives from outside world representatives can combine to take him out of the peer-group society." Gans's analysis implies that if individuals succeed in breaking loose from a more localistic to a more cosmopolitan society, they clearly must have the capacity to do so. Those with cosmopolitan outlooks, life styles, and social networks, in other words, must possess the underlying abilities needed to sustain these. But Gans (who does not seriously consider an underlying genetic or neurobiological dimension in his study) also indicates the possibility that a would-be cosmopolitan may become trapped in a localistic society. So, it is imperative to keep in mind the wide range of individual variation that will likely characterize any population, including the most localistic.

In any case, we propose the following local-cosmopolitan continuum based upon these differing types of patterns of social interaction:

1. the social isolate, who has no informal web of social ties at all, that is, no social network.
2. the neighborhood local, whose social network is largely restricted to an area within walking distance of his or her residence.
3. the community local (Merton's local influential), most of whose social network is interspersed within his or her home community (or distinctive section of a large city).
4. a regional cosmopolitan, most of whose social network is scattered throughout a metropolitan or regional area.
5. an international cosmopolitan, whose social network is distributed widely beyond regional and national borders. (For an elaboration of this continuum see White [1959]).

All of the continua thus far described may now be summarized in relationship to one another as shown in Table 4.1.

The Socio-Cultural Dimension

Merton (1957, p. 393, fn. 7) notes that his local-cosmopolitan typology is based upon "Toennies' well-known distinction between Gemeinschaft (localistic) and Gesellschaft (cosmopolitan)." Too rigid an identification of the members of a small close-knit society as locals and members of a large modern society as cosmopolitans, however, must be resisted. Gans's study of the "urban villagers" indicates that highly localistic communities can exist in Western industrial society.

Mead's study of the Paliau Movement in the Admiralty Islands (1964, ch. 9) reveals that the most isolated primitive society may give rise to a cosmopolitan leader: "With a mind comparable to that of Franklin Roosevelt, Paliau lived in

Table 4.1
Relationships between Space-Time Horizon, Local-Cosmopolitan Orientation, and Patterns of Social Interaction

1.	Space-time horizon (as capacity)	Limited	---------->	Extended
2.	Local-cosmopolitan orientation (as capacity)	Localistic	-------->	Cosmopolitan
		Present oriented	-------->	Future oriented
3.	Patterns of social interaction	Proximity as basis	------->	Mutual interest as basis
		Social Isolate (no social network)	----->	International cosmopolitan (global social network)

a world which consisted of almost illiterate people'' (Mead 1964, p. 229). Mead (1964, p. 228) suggests that had Paliau been born a hundred years earlier, his rise to prominence might not have been possible. Even as a child growing up in a small isolated society, he was, by his own account (Mead 1964, p. 219), resistant to the religious beliefs of his native island: ''I never used to conform properly to any belief.'' Subsequent travel to others of the Admiralty Islands proved disappointing because they observed similar customs.

The Paliau movement, which he founded and led, was an attempt to unify and modernize the Admiralties. According to Mead (1964, p. 288), its program, which Paliau contributed, ''had the character of all great programs in providing a long time perspective and a larger context for the human beings to whom it was addressed.''

Although Paliau's leadership was critical to the movement, its success depended as well on the organizational abilities of the men who became Paliau's lieutenants, ''those who could share his vision'' (Mead 1964, p. 233). To employ Mead's terminology (1964, pp. 54–55; emphasis in the original), Paliau succeeded in forming a ''cluster,'' which Mead sees at the level of cultural micro-evolution ''as the significant *biological* unit'' and which involves ''the specific genetic constitution'' of the particular individuals in interaction with one another. (By 1964, Margaret Mead had formulated an explicitly *bio*cultural framework of analysis). The key point for our purposes is that a cosmopolitan-led cluster emerged out of a localistic society.

One final point: In some cases, members of a community may be simultaneously local *and* cosmopolitan. This case might exist when a group of cosmopolitans, who interact intensively with one another, reside together in the same community. Their orientation and patterns of social interaction would be neither exclusively local or cosmopolitan, but rather include both dimensions. Adler and Bobrow (1956) describe such a community in their study of ''Aurelia,'' an affluent Chicago suburb on the North Shore.

The Political Dimension

If we take the argument presented in chapter 2—that like-minded individuals seek out and socialize with other like-minded individuals—such ensuing social networks will intensify the social and psychological attributes that their members bring to them. This argument will now be applied to local and cosmopolitan social networks, with special emphasis on the resulting political implications.

If cosmopolitan social networks are formed on the basis of mutual interests, we should expect that these interests will be discussed, intensified, and acted on. But cosmopolitan ties may also have a broadening effect, and not merely because they presuppose a broad territorial basis. The more cosmopolitan the individual, the more likely that he or she will encounter a diversity of people, not perhaps representing all walks of life, but at least representing different cultural backgrounds and social achievements. These far-flung contacts should then be educative and focus attention on matters otherwise foreign, if for no other reason than that of having to make conversation.

Charles Chaplin, for example, having risen from a poverty-stricken English background to international film stardom, rubbed shoulders with the likes of Winston Churchill and Mahatma Gandhi. Regarding meeting the latter, Chaplin (1964, p. 342) writes, "The Mahatma had not yet arrived; and as I waited I began to think of what I would say to him." Upon being introduced, Chaplin goes on to comment, "Now came that uneasy, terrifying moment when I should say something astutely intelligent upon a subject I knew little about." Cosmopolitan ties, then, may bring a pressure to be, or at least appear to be, well-informed along a broad variety of topics. This observation receives some support by Maurer's study of confidence men, whom he states are "the elite of the underworld," "the most cosmopolitan of all thieves, for they travel widely and cultivate associations on a rather high cultural level" (Maurer 1974, pp. 144, 146–47). More to the point, the marks cultivated and taken by the con men are typically cosmopolitan in orientation and social interaction and are frequently and deliberately encountered during travel.[1] Con men must be prepared for all levels of conversation. As Maurer (1974, p. 158) points out, "They must be well informed in business and financial matters, have a glib knowledge of society gossip, and enough of an acquaintance with art, literature, and music to give an illusion of culture. I have observed that many of them read widely, and that it is an almost universal habit to run through ten or a dozen newspapers daily in order to keep constantly informed on topics which may come up in conversation."

By virtue of membership in cosmopolitan social networks, then, an individual may have more of an opportunity to pick up information of a political nature than someone who has only local ties. Crenson's comparative case study of six neighborhoods in Baltimore (1978) lends some support to this assertion. Crenson (1978, p. 581) distinguished between "close-knit" and "loose-knit" social networks, the former consisting of people living in the same neighborhood, who know one another, and by and large are friends. The "loose-knit" networks are made up of "cosmopolitan" residents who report that most of their friends live outside the neighborhood and were not acquainted with one another. Nonetheless, it emerged that knowledge concerning local community associations was "most widespread by far among respondents with 'loose-knit' neighborhood friendship networks" (Crenson 1978, p. 581):

> The defining characteristic of a close-knit network is that a person's friends tend to be acquainted with one another. The people with whom he speaks are also in communication with each other. This means that he may hear the same piece of neighborhood news several times over; . . . Gossip may become more memorable through such repetitious communication, but it is not likely to spread very far.
>
> Broader dissemination can be expected through a loose-knit network, in which a person exchanges information with friends who tend not to be acquainted with one another, and therefore gets the news from people who do not share the same information. Instead of

hearing the same news over and over again, like his ''close-knit'' counterpart, he is apt to receive a wider variety of information from more diverse sources and so becomes more generally informed about neighborhood happenings. The result is all the more striking because loose-knit responders reported having fewer acquaintances in their neighborhoods, than did close-knit residents.[2]

Crenson (1978, p. 888) also points out that ''cosmopolitan residents use means other than personal communication for spreading information,'' including newspapers, handbills, and circulars. Locals, by contrast, tend to be restricted to their personal (and parochial) sources of information.

Gans's study of the local West Ender (Boston) supports this latter conclusion. According to Gans (1965, pp. 188–93), West Enders read little and confine their visual media viewing to a few favorite types—action dramas and sports for men, popular musicals and soap operas for women. Peer group conversation similarly ''covers a relatively small number of topics,'' with ''much of the conversation'' being ''devoted to the exchange of reportage—and gossip—about people known to the group.'' (Gans 1965, pp. 77–78).

Consider now how local and cosmopolitan social networks intensify the attributes their members bring to them. Cosmopolitans, with broad space-time horizons and, therefore, the conceptual capacity to direct the course of their lives and perhaps those of others, join others with similar capacities as well as interests. The resulting social network leads to an exchange of information from diverse sources and places, some of which may be inherently political, or at least potentially so in the comparison and contrast over time and space. ''Does such and such a community or country do something better?'' ''Will our present course find us worse off in the next decade because of anticipated (demographic) changes?'' ''Then let us improve or change our present course of action,'' may well be the cosmopolitan response. If this response finds reinforcement in a social network composed of like-minded individuals, then the ''hypertrophy'' that Wilson (1978) describes or the ''synergism'' that Corning (1983) calls attention to comes into play, amplifying tendencies, which if found only in isolated individuals, might never materialize.

Local social networks amplify different tendencies. The extremely localistic social network that Gans (1965) describes produces an atmosphere that is intensely conformist, xenophobic, personalistic, apolitical, and fatalistic. As discussed earlier, the object-oriented individual who is socially and spatially mobile is usually strongly rejected by such a peer-group society.

Between the more extreme local and cosmopolitan poles, however, may exist an intermediate position, which may, in some ways, be represented by Merton's ''community local.'' This individual may be local because of community identification, interest, and social interaction; nonetheless, by embracing the community as a whole rather than being restricted to a single neighborhood, the individual may exhibit cosmopolitan capacities sufficient for political interest and action to emerge, albeit on a local level. The parochial space-time horizon remains, however, and the synergistic effect of the social network no doubt intensifies the provincialism.

Case Studies: Response to Urban Renewal

Two communities, at polar ends of the local-cosmopolitan continuum, illustrate likely differences in political activities by a focus on their differential response to urban redevelopment. In this discussion, keep in mind that intermediate positions, including the "community local" would not necessarily inhibit political activity, even if they influenced political orientation.

Localistic Community. For a localistic community, we again rely on the Boston West End, described by Gans (1965), and its response to the urban redevelopment during the 1950s that would clear the area and relocate its residents. Although the West Enders were strongly attached to their physical neighborhood and opposed to redevelopment, they were unable to resist it. The most important reason for this failure, according to Gans (1965, p. 296), was "the inability of the West Enders to organize in their own behalf." Such political organization suggests capacities that most West Enders lacked: the ability to transcend their intensely localistic peer group society, to gather and digest information from what Gans refers to as the "Outside World," and to think and plan ahead. The lack of these attributes dictated a passive, fatalistic outlook incompatible with effective political action. This situation, which Gans (1965, ch. 13) elaborates, meant that "the oncoming redevelopment had little impact on the lives of most West Enders until the very end."

Cosmopolitan Community. A different response to urban renewal came from the residents of Hyde Park-Kenwood in Chicago, the home of the University of Chicago. Rossi and Dentler (1961, p. 38) bring out both the cosmopolitan and the organizational character of the community:

> Given its institutions and the high socio-economic level of its residents, it is not surprising to find that Hyde Park-Kenwood has always supported a dense network of organization. We would expect that any upper-middle-class neighborhood would be well organized. The presence in the area of the University of Chicago and other educational institutions strengthened this tendency by bringing to the area faculty families with more than usual interest in cultural and political affairs. In turn, the presence of this intellectually oriented component attracted to the area a population that shared these interests.

The University of Chicago was the prime mover in planning for urban redevelopment in the community in an effort to safeguard its upper-middle-class character against lower-class encroachments. Nonetheless, the civic association that the residents formed, the Hyde Park-Kenwood Community Conference, played a key role in the renewal process: "The Conference stimulated citizen participation in the planning process in order . . . that the planners could produce a plan more closely attuned to the felt needs and desires of residents. Widespread citizen participation produced equally widespread public acceptance of the plan" (Rossi and Dentler 1961, p. 5).

The contrast with the West End urban development is apparent: the almost total absence of organization in the community in the once case, and the widespread organization in the other. In addition, the outcome of the political process was markedly different: the West Enders found their neighborhood and their homes razed; the Hyde Park-Kenwood residents were, from their perspective, "improving" a neighborhood where they would continue to live. As Wilson (1967, pp. 291–92) explains: "The upper-middle-class professors, housewives, and business and professional men (both white and Negro) who made up the bulk of the Conference were mostly people who were going to remain in the community and whose peace, security, cultural life, and property values would probably be enhanced by a successful renewal plan. The persons who were to be moved out of the community and whose apartments and homes were to be torn down were usually lower-income Negroes who, with very few exceptions, were not part of the Community Conference."

Wilson also notes, however, that the role of the Conference was not limited to one of advocating a position of narrow self-interest. Rather, its members were firmly committed to the idea of a multiracial community and some of them to a multiclass community. "Time and again the Conference, or leading members of it, pressed the city to provide middle- and low-income public housing in the renewal area in part to accommodate persons who would be displaced by demolition" (Wilson 1967, p. 292).

Wilson (1967, p. 292) concludes:

> This broader consideration was equally as important as the narrower one in explaining the positive and constructive role of the Conference. The organization was made up to a large degree of persons who attached a high value to community-wide and neighborhood-wide goals, even (in some cases) when attaining those goals entailed a sacrifice in personal, material satisfactions. They are people who partake to an extraordinary extent of what Edward C. Banfield and I have called . . . the "community-regarding" or "public-regarding" political ethos. This ethos, which is most likely to be found among citizens who rank high in income, education, or both, is based on an enlarged view of the community and a sense of obligation toward it. People who display it are likely to have a propensity for looking at and making policy for the community "as a whole" and to have a high sense of personal efficacy, a long time perspective, a general familiarity with and confidence in city-wide institutions, and a cosmopolitan orientation toward life. In addition, they are likely to possess a disproportionate share of organizational skills and resources.

Wilson (1967, p. 293) goes on to observe that for most renewal areas many of the inhabitants "are the opposite in almost every respect from the cosmopolitan elite of Hyde Park-Kenwood": "Such people are more likely to have a limited time-perspective, a greater difficulty in abstracting from concrete experience, an unfamiliarity with and lack of confidence in city-wide institutions, a preoccupation with the personal and immediate, and few (if any) attachments to organizations of any kind, with the possible exception of churches." Wilson proceeds to footnote, for purposes of illustration, Gans's case study of the West End.

Thus, Wilson's dual conclusion may serve as our own, keeping in mind that, as in all cases of human life, there is an important genetic and neurobiological dimension that should be viewed, as should the environmental dimension, within a populational perspective.

NOTES

1. A note on travel as a cosmopolitan trait is in order. First, travel as such is neither cosmopolitan nor broadening if an individual lacks the capacity for an extensive space-time horizon. Just as a drafted foot soldier might travel widely and learn little, a twelfth-century crusader might "travel much and yet see little" and fail to be "intellectually changed by his crusading" (E. J. Passant; cited in Haskins 1961, p. 15). On the other hand, a true cosmopolitan may stay in one place for a lifetime or a good length of time and still read widely and meet anybody who is anyone who passes through. Such seems to be the case with Immanuel Kant, who never journeyed far from Königsberg in East Prussia, or Albert Schweitzer, during his largely unbroken quarter-century sojourn in Lambaréné in what then was the Gabon province of French Equatorial Africa.

2. However valid Crenson's findings may be for Baltimore, they may not be generalizable to other communities. At least two possible qualifications come to mind. First, in a community like "Aurelia" (Adler and Bobrow 1956), where cosmopolitans are simultaneously local, Crenson's distinctions may not readily apply in the first place. Second, in a smaller community like Merton's "Rovere" (Merton 1957, ch. X), where cosmopolitans are considerably outnumbered by community locals and are in any case attracted to cosmopolitan centers like New York, their interest and involvement in local affairs is likely to be quite limited, especially in relation to the more numerous and active locals.

Chapter 5

The Founding Fathers as Cosmopolitans: The Neuropolitics of the Constitution

> Mr. Gov. Morris . . . came here as a representative of America; he flattered himself he came here in some degree as a Representative of the whole human race: for the whole human race will be affected by the proceedings of this convention. He wished gentlemen to extend their views beyond the present moment of time; beyond the narrow limits of place from which they derive their political origin.
>
> —James Madison, *Notes of Debates in the Federal Constitution*

> In pursuing this inquiry, we must bear in mind that we are not to confine our view to the present period, but to look forward to remote futurity.
>
> —Alexander Hamilton, *Federalist No. 34*

> Let him reflect that the object upon which he is to decide is not a particular interest of the community, but the very existence of the nation.
>
> —Alexander Hamilton, *Federalist No. 85*

Locals versus Cosmopolitans: that is how several recent historical accounts of the fight over the ratification of the U.S. Constitution describe the main battle lines. In this chapter I basically accept this description; my purpose is to add a biological dimension totally absent from these accounts and to indicate its importance.

FEDERALISTS AND ANTIFEDERALISTS AS LOCALS AND COSMOPOLITANS

Chief of the historians describing the struggle between Federalists and Antifederalists as a local-cosmopolitan split are Main (1961, 1973) and Wood (1969, 1987). Main (1973, p. 22) distinguishes between "cosmopolitan" and "localist" as referring to two opposing world views:

> At one pole is the man of broad outlook, usually urban, urbane, and well educated, who has traveled widely and has had extensive contacts with the world because of his occupation, the offices he has held, or his interests. Jefferson and Franklin were cosmopolitan men. At the opposite pole stands the man of narrow horizons—most often rural and sparsely educated—whose experience is limited to his own neighborhood; he is provincially parochial.

Main (1973, p. 358) applies this distinction to the Federalists and Antifederalists. Classifying members of state legislatures just before the ratification of the Constitution as either localists or cosmopolitans, he finds that the localists were overwhelmingly (84 percent) Antifederalists and the cosmopolitans just as overwhelmingly (85 percent) Federalist. Main (1973, p. 358) notes that each side was united by one central conviction: "The Antifederalists defended local autonomy, while the Federalists sought centralization" and that this "basic ideological difference obviously resembles, if it does not in fact result from, the localist-cosmopolitan dichotomy."

Wood (1969, p. 520) also refers to the Antifederalists "as 'localists,' fearful of distant governmental, even representational, authority." The Federalists, on the other hand, deplored the parochial outlooks of both constituents and their state legislators under the Articles of Confederation. "Too many of the American people could not see beyond their own pocketbooks or their own neighborhoods. 'Individuals of extended views, and of national pride,' said Madison, might be able to bring public proceedings to an enlightened cosmopolitan standard" (Wood 1987, p. 75). Wood (1987, p. 76) also states that "many of the delegates to the Philadelphia Convention were so ready to accept Madison's radical Virginia Plan and its proposed national authority to veto all state laws precisely because they shared his deep disgust with the localist and interest-ridden politics of the state legislators."

The cosmopolitan skills and organization of the Federalists were distinct advantages. Conversely, the Antifederalists were handicapped by their lack of these abilities. "Madison's designation of the Massachusetts Anti-federalists," according to Wood (1969, p. 486), "was applicable to nearly all the states: 'There was not a single character capable of uniting their wills or directing their measures. . . . They had no plan whatever.' "

Wood (1969, p. 486) concludes: "But the Antifederalists were not simply poorer politicians than the Federalists; they were actually different kinds of politicians. Too many of them were state-centered men with local interests and loyalties only, politicians without influence and connections, and ultimately politicians without social and intellectual confidence.''

The benefits the Federalists received from these liabilities on the part of their opposition were enhanced by their own strengths. Not least of their assets were their superior cosmopolitan-based social ties: ''The ability of the Federalists to outmaneuver their opponents was due in part to superior organization. The Antifederalists had not been able to unite, even within a particular state, in order to concert their efforts, until the creation of the Federal Republic Committee in New York. Everywhere they were too late.'' The Federalists, on the other hand, ''were consistently ahead of the game. . . . Certainly their correspondence shows far better cooperation'' (Main 1961, p. 252). Thus, Rossiter (1966, p. 106) describes Madison as ''the stoutest of Virginia nationalists who struck up friendships with like-minded delegates that were to prove immensely helpful in his campaign for a stronger national government.''

Two qualifications are in order. First, we do not assume that a cosmopolitan outlook is necessarily a superior one. In retrospect, it is easy to make such an implicit assumption when considering the ratification of the Constitution. Many observers during Watergate and Iran-Contra were reminded of the legitimacy of the concerns of the Antifederalists. Plato and Aristotle, those two most cosmic of thinkers, preferred small political communities to unwieldy heterogeneous Babylonia—whose liabilities as well as any strengths are shared by our large American republic.

Second, we reiterate the presence of an explicit biological dimension, which is missing in the historical literature. Main (1973, p. 32), for example, holds that the opposing localist and cosmopolitan world views ''are based on the contrasting experience of men''—in short, an environmentalist premise. But as we have seen in chapter 4, a local by nature may travel the world without appreciably broadening his or her horizon. On the other hand, a cosmopolitan by nature, such as Immanuel Kant may never travel beyond his or her small hometown and yet be capable of the most abstract weltanschauung. (Neither Main nor Wood, incidentally, specifically refer to the local/cosmopolitan distinction made in contemporary sociology).

Within this historical context, the terms ''cosmopolitans by nature'' and ''locals by nature,'' need to be elaborated on and compared to the terms ''cosmopolitans by convention'' and ''locals by choice.''

Cosmopolitans by nature. Cosmopolitan individuals by nature have the genetically based capacity for the development of a broad space-time horizon and together with a favorable environment and at times despite a negative one (cf. Alexander Hamilton), will become cosmopolitan. In the case of the most prominent and influential Federalists at the Constitutional Convention in 1787—Benjamin Franklin, George Washington, Alexander Hamilton and James Madison—all could be considered cosmopolitans by nature, with critical significance for both the formulation and adoption of the new national government.

Cosmopolitans by convention. Not everyone who favored the new constitution was a cosmopolitan by nature. Indeed, if the adoption of the Constitution depended upon this condition, it would have been overwhelmingly defeated.

It is clear that many "common people," likely neighborhood locals, also favored the new government. Such individuals would be cosmopolitans by convention—those who do not have a natural cosmopolitan tendency but tend to follow the lead of those who do. Most likely, such individuals are neighborhood locals who are appealed to by both community locals and by cosmopolitans. Lacking any broader political plans of their own, they are dependent on others for their formulation, but they may choose on the basis of the prestige and prominence of the leader who proposed them; this possibility may favor more cosmopolitan leaders.

In the case of the struggle over the Constitution, Main (1961, p. 253) notes that "an advantage of the Federalists was the superior prestige of some of their leaders. In many states this advantage was very great." The Antifederalists simply could not compete with a Benjamin Franklin or George Washington.

Locals by choice. We have already noted that Plato and Aristotle rejected the prospect of governance in a large state, preferring the more intimate polis. By the same token, cosmopolitan thinkers may favor local government.

A number of the Antifederalists fall into this category. Wood (1987, p. 43) observes that "some of the prominent Antifederalist leaders, such as Elbridge Gerry, George Mason, and Richard Henry Lee, scarcely represented, either socially or emotionally, the main thrust of Antifederalism": "Such aristocratic leaders were socially indistinguishable from the Federalist spokesman and often were as fearful of the exercises of democracy in the state legislatures as the Federalists."

Their cosmopolitan character not only showed in its disapproval of parochial excesses, but in its recognition of the need of some centralizing reforms. Some of the Antifederalists, writes Jensen (1958, pp. 424–25), "were in fact as narrow and provincial as their opponents said they were. However, the best of them agreed that the central government needed more power, but they wanted most powers given so as not to alter the basic character of the Articles of Confederation" (cf. however, Elkins and McKitrick 1978).

Thomas Jefferson, in 1787 and later, came close to being a cosmopolitan who was a local by choice. Jefferson, who was abroad in Paris at the time, supported the new constitution, but he shared the Antifederalist concerns over the absence of a bill of rights and over undue executive power. In any event, the locals by choice did not provide the main thrust of the Antifederalist opposition, as did the locals by nature.

Locals by nature. The bulk of Antifederalist support came from people genuinely attached to the local political institutions that were under attack by the Federalists. Unlike the cosmopolitans by nature or those who were locals by choice, they had no real capacity to transcend their social and political milieu. But unlike many neighborhood locals who were perhaps cosmopolitans by convention and relatively stationary in life, they had the drive and natural talent to move up to social and political positions of local leadership, which they valued, identified with, and wished to protect. A typical example of such an Antifederalist was Pennsylvania's William Findley.

Two brief points need to be made before moving on to a consideration of the leading Federalists. First, community locals are not thought to be lacking intellect, but they are more likely to possess a shrewd rather than reflective intelligence. Second, insofar as it is generally true that the more limited one's space-time horizon, the more closely one is tied to the environment, so these locals are more likely to be tied to their specific environmental circumstances—compared, for example, to the cosmopolitans who are locals by choice.

THE FOUNDING FATHERS AS COSMOPOLITANS

Benjamin Franklin

There can be no doubt that Benjamin Franklin was a cosmopolitan. But it is interesting to see how this characteristic may have had an underlying genetic basis and how in turn this may have had a direct bearing on Franklin's political views.

Franklin observed that he "emerged from the poverty and obscurity in which I was born and bred, to a state of affluence and some degree of reputation in the world" (Franklin 1966, p. 6). When he runs away from an apprenticeship in his elder brother's print shop in Boston he does so as "a boy of but 17, without the least recommendation to, or knowledge of any person in the place, and with very little money in my pocket" (Franklin 1966, p. 28).

Franklin's arrival and swift rise in Philadelphia are well known. Following is Carl van Doren's assessment (1938, p. 118) of that rise:

> The industry and frugality about which he talked cannot account for his swift progress. That he talked about them so much makes clear that they came less from his nature than from his discipline. Other men besides Franklin in Philadelphia were as industrious and frugal as he. No other man had a mind so gracious and ingenious and incessant, so able at once to persuade and to charm. It was Franklin's luck that it was easier for him to be outstanding in Philadelphia than it would have been in London. But from the first he was outstanding, and he throve by the exercise of natural gifts of which Poor Richard could not tell the secrets.

Franklin's IQ up through age 25 has been estimated at 145 by a panel of three psychologist judges, including Lewis Terman (Cox 1926). This represents actually a kind of minimal IQ, since it is based on the response to the following question: Given what we know of Franklin's (or whoever's) early creative work—still extant plays, poetry, music, etc.—what minimal IQ would be necessary in order to account for it? Thus this estimated IQ is not quite so outlandish as it might at first appear, and in any event it no doubt underestimates actual intelligence if it does anything at all. Assuming his IQ were between 140 and 160, Franklin would be equalled or excelled by anywhere from 1 to 70 out of a population of 10,000 (Cox 1926, ch. 4). The judges in fact estimate that their "minimal" IQ of 145 is more likely equal to a "true" IQ of 165.

Like van Doren, I assume that "natural gifts, as well as environmental influence lie behind Franklin's high intelligence—that is, that both heredity and environment make an important contribution. Franklin (Franklin 1966, pp. 11, 12) refers to his "early readiness in learning to read (which must have been very early, as I do not remember when I could not read)"; and van Doren notes that Franklin "was almost as wholly self-taught as if he had never gone to school (as it was, he spent only two years in formal schooling). The most insatiable and acquisitive young mind in America was on the hunt for knowledge and would have found it in a desert."

Such an inquisitive, wide-ranging mind embraced the widest space-time horizon. Wright (1986, p. 13) states:

> What makes Franklin so compelling to study and so original is his almost total freedom from the limits of his own environment. He made the whole western world his own parish. Born poor in colonial Boston, he was at ease in Quaker Philadelphia, in royal London, and in elegant Paris; as much at home in St. Andrews as in Passy, or in Bethlehem and Easton on the Pennsylvania wilderness frontier. The reason is clear. He was far in advance of his time, not merely as editor turned scientist turned statesman, or from the prescience of so much of his writing, but because one is aware of his own awareness of challenges and of the immense potential of the human mind when left free to speculate, to question, and to experiment. . . . [He] knew no frontier. Four months before his death he wrote his English friend David Hartley, "God grant that not only the Love of Liberty but a thorough knowledge of the Rights of Man may pervade all the Nations of the Earth so that a Philosopher may set his foot anywhere on its surface and say, 'This is my country.' "

"Of the fathers of his country," Wright (1986, p. 359) concludes, Franklin "was at once the most American and yet the most cosmopolitan, the most prescient for the future, the new man."

This cosmopolitan outlook had its political and economic dimension: Franklin "advocated colonial union long before it became politically possible, union that should be not only political but economic, built on a sound federal basis of postal roads, corresponding philosophical societies, and a strong central government, inside or outside the Empire" (Wright 1986, p. 350). In 1754, Franklin proposed the Albany Plan, based upon his earlier 1751 "Plan of union" which envisioned a union of all the British colonies (except Georgia and Nova Scotia) under a president general appointed and paid by the crown, together with a grand council elected by the colonial legislatures.

The Albany Plan of Union was rejected by both the home government and the colonies. Nonetheless, the plan was, according to Morris (1958, p. 26), "a grand design of statesmanship, the kind that is envisioned perhaps not more than once a century": "Ultimate recognition of the practicality of the Albany Plan was attained at the Constitutional Convention. If you substitute a president for a president-general and add a second house, you will find that in substance the Albany Plan was embodied in the federal Constitution" (Morris 1958, p. 38).

George Washington

If Washington were indeed the ''indispensable man'' for the success of the American revolution and of the later national constitutional government, it was not due to his character alone. Yet ''[b]y common consent his intellectual talents were limited'' (Borden 1969, p. 1).

First when examining his childhood, Washington's background was not so advantageous as is commonly thought:

> No American is more completely misunderstood than George Washington. He is generally believed to have been, by birth and training, a rich, conservative, British-oriented Virginia aristocrat. As a matter of fact, he was, for the environment in which he moved, poor during his young manhood. He never set foot in England or, indeed, any part of Europe. . . .
>
> No other President of the United States before Andrew Jackson was as much shaped by the wilderness as Washington, and he had less formal education than did Jackson, than Lincoln even. Both Jackson and Lincoln studied law, while Washington's total schooling hardly went beyond what we should consider the elementary grades. (Flexner 1984, p. 1)

Between such unpromising beginnings and the generally held low estimate of his intellect, ascribing a sophisticated cosmopolitan space-time horizon to Washington appears unjustified. But there are three reasons why Washington's intellect may be undervalued:

First, the emphasis on Washington's character overshadows his intellect. This emphasis appears in the standard treatments of Washington's life, including Douglas Freeman's monumental biography. Yet Freeman (1948, vol. II, p. 382) also notes Washington's ambition and his unusually rapid rise: ''His rise had been much more rapid than most of his contemporaries.'' In 1749, at age seventeen, Washington was appointed the official surveyor for Culpeper County in Virginia; at twenty-two he was appointed lieutenant colonel of the Virginia militia; and at twenty-six he was elected to the Virginia House of Burgess. Freeman compares this record favorably with that of John Adams, Samuel Adams, John Hancock, James Otis, Patrick Henry and even Benjamin Franklin.

Such a rise cannot be due only to birth, luck, and character. According to *Genetic Studies in Genius* (Cox 1926, p. 345), Washington's minimal IQ at age twenty-six was 135. This estimate would place Washington equal to or superior to such political or military leaders as James Madison (135), John Churchill (125), Giuseppe Garibaldi (125) and Joachim Murat (120). Again, these estimates are imperfect, based upon incomplete historical records, but, if anything, they underestimate actual intelligence. Thus, the authors estimate the likely or ''corrected'' IQ for 135 as 155, which would of course place Washington well into the upper one percent of the population.

Second, the underestimation of Washington's intellect results from unfavorable comparisons with his political contemporaries. On the basis of his actual accomplishments, Washington can hold his own, since he was unanimously named commander in chief of the American Forces during the Revolutionary War, unanimously named president of the Constitutional Convention in Philadelphia, unanimously elected first president of the United States (and later to be unanimously reelected). Nonetheless, surrounded by the likes of Benjamin Franklin, John Adams, Thomas Jefferson, Alexander Hamilton, and James Madison, Washington appears eclipsed by their brilliance.

Such comparison, however, is not necessarily demeaning.[1] Washington's peers were indeed brilliant. Examine, for example, Jefferson's assessment (cited in Borden, 1969, p. 105) of Washington's intellect: "His mind was great and powerful, without being of the very first order; his penetration strong, though not so acute as that of a Newton, Bacon, or Locke." If compliments can be backhanded so slights can be complimentary, and Jefferson's slighting comparison is surely that. What recent American president can even be mentioned in the same breath with Newton, Bacon or Locke, without the comparison seeming ludicrous?

In addition, Washington lacked the surface brilliance of such contemporaries as Hamilton. Washington would be bound to suffer by comparison, in much the same way that someone personally meeting Marx and Engels, or Einstein and Van Neuman for the first time might judge the latter of two coupled names more brilliant (Engels and Van Neuman *were* brilliant) than the former. The former—Marx and Einstein—had, however, deeper minds. This observation is not made to place Washington in the same company as Marx and Einstein as a heavy-weight thinker, but only to suggest that although Washington's intellect lacked surface brilliance, it was nonetheless substantial in its own way.

Third, in conjunction with Washington's prudent, judicious character, his mind was slow to move and so could appear to be dull. Consider again Jefferson's assessment (cited in Borden 1969, p. 105): "as far as he saw, no judgment was ever sounder. [His mind] was slow in operation, being little aided by invention or imagination, but sure in conclusion. Hence the common remark of his officers, of the advantage he derived from councils of war, where hearing all suggestions, he selected whatever was best; and certainly no General ever planned his battles more judiciously."

What Jefferson seems to say is that Washington was slow to decide but invariably decided well. In his *Presidential Greatness*, Thomas Bailey (cited in Borden 1969, pp. 142–43), places Washington "at the very top," in part because "the absence-of-blunder test leaves Washington with a uniquely high mark": "Although his every move could be deemed a potential precedent binding generations unborn, his foot did not slip once. He made no major mistakes, something that cannot be said of any of his successors who served long enough to make a mistake." Bailey also extends this judgment to Washington's military career: "The Irish-British historian W.E.H. Lecky, writing of Washington's leadership during the Revolution, concluded that 'of all the great men in history,' the august Virginian was the most "invariably judicious.' ""

An indicator of Washington's intelligent and judicious leadership ability was his capacity to spot and to appoint talent, from the young Alexander Hamilton during the Revolution and later as Secretary of Treasury to Thomas Jefferson as Secretary of State.

The absence of any major misjudgment on Washington's part during his long and illustrious public career cannot be merely the result of luck and good character. Flexner's verdict (1967, p. 552) appears justified: "Again and again to discover, when immersed in the welter of troubled times, the best routes ahead is a towering intellectual achievement."

Washington's intellectual horizon was broad. "It should be the highest ambition of every American," Washington wrote, "to extend his views beyond himself, and to bear in mind that this conduct will not only affect himself, his country, and his immediate posterity; but that its influence may be co-extensive with the world, and stamp political happiness or misery on ages yet unborn" (cited in Flexner 1967, p. 552). Comparing him to Patrick Henry and older Virginia leaders, Perry (1984, p. 133) states that Washington "epitomized a new style of moderate, enlightened leadership": "Like his predecessors, he was renowned for his horsemanship and excellent conversation. But he was well read in the philosophy of an enlightened age."

Washington's broad cosmopolitan perspective revealed itself during the course of the American Revolution when he felt the deep need for a stronger national government. At the conclusion of the war in 1783, he wrote a circular letter, later known as "Washington's legacy," in which he urged that the Articles of Confederation be extended to create a central government adequate to obvious needs. Privately, according to Flexner (1984, p. 180), Washington had already been urging a constitutional convention that would not take place for another five years. When soon after the Revolution he undertook the leadership of the Potomac land project that would lead to a series of political meetings culminating in the Philadelphia constitutional convention in 1787, Madison commented on Washington's role that it "shows that a mind like his, palpable of great views and which has long been occupied with them, cannot bear a vacancy" (cited in Flexner 1984, p. 196).

In 1786, when the Philadelphia convention for which he would serve as president was set, Washington admonished in a letter to Madison (cited in Brant 1948, p. 391): "Let us look to our national character, and to things beyond the present period." In short, Washington's broad space-time horizon was clearly in evidence in his continuing concern for events beyond his Virginia estate and their impact on a long-term future.

Alexander Hamilton

The lowly origins, brilliant mind, and rapid rise of Alexander Hamilton are well known, as well as his early championing of a strong national union. We will briefly review each in our context.

When Alexander's father, James Hamilton, abandoned his mother in the West Indian island of St. Croix, he left the family destitute. At the age of eleven, Alexander became a clerk in the counting house of Cruger and Beckman. Later in the *Federalist Papers*, Hamilton wrote: "There are strong minds in every walk of life that will rise superior to the disadvantages of situation and will command the tribute due to their merit, not only from the classes to which they particularly belong, but from the society in general." Miller (1959, p. 4), notes Hamilton's reference to his own situation here, including his "keen awareness" of the "superiority" of his talent. "One is constantly prompted," writes Mitchell (1957, p. 207), "to lay Hamilton's superior grasp and power of statement to inborn gift and let it go at that. Beyond a point this must be the plea, for his best work was of a different order of excellence from the offerings of most of those about him."

According to *Genetic Studies in Genius* (Cox 1926, p. 409), Hamilton's minimal IQ at age 26 was 140. At 15 he migrated to New York and proceeded to attend a grammar school in New Jersey for one year. At 16 he passed the entrance examination to Princeton, but since he was refused permission to complete the curriculum in three years, he entered King's College (Columbia), where he did so.

At 18 he was captain of a New York company of artillery in the Revolutionary War. Two years later, in 1777 at age 20, he was appointed Washington's secretary and aide. At 25 in 1781, he was elected to Congress.

As early as Valley Forge Hamilton launched his first known advocacy of the superiority of the Federal system of government: "How can we hope for success in our European negotiations if the nations of Europe have no confidence in the wisdom and vigor of the great Continental Government?" (cited in Flexner 1978, p. 208). Hamilton thereafter consistently and forcefully advocated a stronger national union.

Hamilton "had little of the sectional and local attachment such as formed a prominent ingredient in most of his party colleagues of New England and with men of opposite persuasion in the South. To a singular degree he was a national figure" (Mitchell 1957, p. 379). His space-time orientation was clearly cosmopolitan. Hamilton did as much, Flexner (1978, p. 436) concludes, "as any man—with the possible exception of Washington—to complete the Revolution by bringing the states together into a unified nation."

James Madison

Precisely because of his "intellectual powers," Madison, according to his major biographer, Irving Brant (1941, pp 39–40), has come down through his previous biographers "as a colorless, detached, passionless intellect." Brant is concerned with humanizing Madison, but for our purposes it is well to remember that Brant, along with the other biographers, does not deny the power of his intellect. Brant (1941, p. 77) refers to his "inquiring" mind, "swift in the development of original views."

Madison's cosmopolitan outlook was nourished at Princeton, which he attended following a comfortable childhood in Virginia as the son of a planter:

> At Princeton for the first time in his life Madison associated with young men from New England, New York, Pennsylvania and New Jersey—the middle and northern colonies of British America—and from Maryland, the Carolinas and Georgia. From the standpoint of student contacts and resulting outlook, it was a drastic change from anything he had known. The effect, reinforcing native tendencies, was to free him from provincial bias, and thus to set him apart from many of his fellows in Virginia during the formative period of state and national life—apart even from Jefferson—with incalculable results upon his contribution to American government. (Brant 1941, pp. 72–73)

But if the Princeton environment enlarged Madison, it was nonetheless "reinforcing native tendencies." Adair (1974, p. 125) writes that Madison, "besides inherited wealth and position" had "an advantage far more important—a first-rate brain":

> Although neither his father, nor his brothers, nor any other members of his immediate family or their descendants ever exhibited any particular intellectual distinction, James Jr., through the mysterious alchemy of the genes, was endowed with a capacity for extraordinary intellectual accomplishment.

According to *Genetic Studies in Genius* (Cox 1926, p. 293), Madison's minimal IQ at age 26 was 135. He had compressed a four-year course of study at Princeton into little more than two years. Princeton, under the presidency of John Witherspoon, encouraged an intellect that was already independently receptive.

> For a person of his mental traits, the atmosphere he encountered at Princeton was favorable to development. Under Dr. Witherspoon's guidance, and transcending it, James Madison entered on a course of study which carried out at extended intervals for fifteen years after he left college, was to make him the best-informed man in America on the principles of government and the keenest analyst the world had yet produced of the relationship between government and economics. (Brant 1941, p. 78)

By 1774, at the age of 23, Madison was already caught up in the revolutionary fervor of the American colonists. He was, moreover, placing it within the broad context of an American—not Virginian—nationalism:

> [A] vibrant continental spirit runs throughout his letters of this period. He described the militant patriotism of the Virginia delegates as "American ardor." The object of their devotion is "American freedom." When he speaks of "our part," "our adversaries," "our secret enemies," he speaks in terms of America not of Virginia. (Brant 1941, pp. 144–45)

Elected to the Continental Congress in 1779, at the age of 28, Madison was among the first to advocate additional powers be given to the Congress when the government under the Articles of Confederation was near collapse due to the states' failure to meet congressional requisitions for the federal treasury. "Through the 1780's," writes Meyers (1981, p. 47), "Madison continued in the vanguard of the movement to strengthen federal authority over commerce, revenue, and foreign relations."

Thus Brant, who subtitles the second volume of his biography of James Madison, "The Nationalist" (Brant 1948, p. 398–99), points out that "the record of his early political nationalism is unmistakably clear" and that his "persisting loyalty to America in general, when state loyalties were swelling, did indeed set him apart from most of his fellows." Finally, as Brant (1948, pp. 398–99) notes, "The continentalism displayed by Madison would be important, even if unique, because of its bearing on his own contribution to American constitutional development." The "father of the Constitution" provided the basis for its formulation as the author of the Virginia Plan, which would clearly shift sovereignty from the individual states to the new national government. In brief, Madison's ample capacity to formulate a cosmopolitan space-time horizon was an indispensable condition for the critical role that he was to play in the formulation of the American Constitution.

We see, then, that these four Federalist leaders seem clearly to be at least national cosmopolitans and, in the case of Benjamin Franklin, an international cosmopolitan. (I would conjecture that some of the more prominent Antifederalist leaders, such as Patrick Henry, veer toward the sectional cosmopolitan category; that is, they are relatively less cosmopolitan.)

In any event, it is my contention that Franklin, Washington, Hamilton, and Madison are cosmopolitans by nature; and that this fact plays a critical historical role as a precondition for their key involvement in constitutional politics. If these four leaders had never been born, or if they had died during infancy, American history might have taken a different turn. This possibility is equally open if all four had been born with only average potential befitting a neighborhood local.

LOCALS BY NATURE

"Far more representative" than the cosmopolitan likes of Elbridge Gerry, George Mason, and Richard Henry Lee "of the populist opposition to the 'aristocracy' of the Federalists was someone like William Findley—that pugnacious Scotch-Irishman from western Pennsylvania" (Wood 1987, p. 93). Keeping in mind that no "typical" representative literally exists, let us nonetheless take a closer look at Findley. Born in Northern Ireland in 1741, he had little formal education, but, according to *The Dictionary of American Biography* (VI, p. 385), "a strong intellect, which he cultivated by reading." He migrated to America in 1763, and worked several years as a weaver before purchasing a farm. Findley entered the American army during the Revolution, rising to the rank of captain. After the war he migrated to western Pennsylvania and entered public life, serving, among other things, as an assemblyman and a delegate to the state constitutional convention of 1789–90.

As an Antifederalist, Findley vigorously opposed the ratification of the Constitution, but, ironically, in 1791 was elected to the House of Representatives where he served continuously until 1817, except for four years (from 1799–1803) during which he served in the state senate. At first (but not later on), Findley encouraged resistance to the national government during the Whiskey Rebellion of 1794.

Wood (1987) explicitly notes the similarities and difference between the careers of William Findley and Benjamin Franklin. Both came from modest beginnings; both had little schooling; both migrated at an early age; both began with a skilled trade (weaving and printing); and both rose rapidly (especially Franklin). But while Franklin became a man of the world, Findley never quite transcended the Pennsylvania frontier.

According to the *Dictionary of American Biography* (VI, p. 386), "Throughout his public career Findley was a faithful guardian of the interests of the frontiersmen, who were his associates and his friends." One of the methods used by the radical party in Pennsylvania—the Constitutionalists—to rally the frontier regions to its support was to use "stump speakers like William Findley who would talk to the settlers in their own language" (Brunhouse 1942, p. 5).

Main (1973, pp. 207, 437) identifies the Pennsylvania Constitutionalists and hence Findley with his "localists"; but he might also be classified in our terms as a sectional cosmopolitan, or at least as something a bit more than a community local. "Because of his large personal acquaintance and his inherent ability he became a formidable factor in sharing public opinion not only in western Pennsylvania but throughout the state" (*Dictionary of American Biography* VI, p. 385). The critical fact is that he was not a national cosmopolitan and certainly not an international one like Franklin was.

It is also instructive to compare Findley with Hamilton. Both were of modest origins, migrated at an early age to a new country, and rose rapidly, though much more so in Hamilton's case. Like Franklin but unlike Findley, Hamilton chose an urban setting (New York) and fashioned a cosmopolitan identity. Now Mitchell (1957, pp. 107–8), while allowing for Hamilton's "inborn gift," asserts that "we can go far in accounting for the unaccountable" and proceeds to argue that Hamilton's strong nationalism was the result of his having "had no birthright": "He was the convert who is more zealous than the inheritor of a faith." Similarly, Morris (1957, p. 60) states that "perhaps because he knew no native allegiance to any of the original states, Hamilton was among the foremost figures of the Revolution to think in continental term." But neither had Findley any "native allegiance." Did that make him a nationalist?

There are no intelligence estimates for Findley. He was clearly able: "Findley displayed early in life that ability for leadership which was for him a long and creditable career in public service. Indeed, he was a consummate politician" (*Dictionary of American Biography* VI, p. 385). Findley himself (1881, p. 441) was to observe that he had "from an early period of life been accustomed" to "reading newspapers and Geography" and that when he moved to his adopted country the acquaintances that he made first "were generally the most intelligent and respectable in the place." Ferguson (1938, p. 40) notes that though he was "of the people, a fact that endeared him to the Westmoreland pioneer" and "fought for the interests of his region," he was not, "as his writings attest, the 'clodhopper' that his enemies portrayed him to us." Finally McMaster and Stone (1970, p. 729) write that he "was a shrewd politician" who "was as forcible a writer as a speaker, and the newspapers of the day contained many political articles from his pen." But just as clearly, extending Wood's comparison (1987) with Franklin, he lacked the abstract, reflective intelligence that would have permitted him to transcend his frontier environment.

LARGER CONTEXT

In this concluding section, I will consider four alternative analyses of the politics of the Constitution: (1) the economic interpretation of Charles Beard; (2) the sectional analysis of Orin Libby, together with the "social interpretation" of Lee Benson; (3) the self-selection thesis of Elkins and McKitrick; and finally (4) a Madisonian understanding.

An Economic Interpretation of the Constitution of the United States

When Charles A. Beard's book, *An Economic Interpretation of the Constitution of the United States*, was first published in 1913 it sparked intense controversy that has not completely abated. It is beyond the scope of this paper to enter this fray (cf. Brown 1956; Benson 1960; and McDonald 1963); I will simply discuss Beard within the present context.

Beard's most general contention (Beard 1961) is that the Constitution cannot be understood without a consideration of economic factors. In his introduction to the 1935 edition (Beard 1961, pp. v-xx), though, Beard is at pains to indicate that his is not an economic determinism. More specifically, and here is where controversy begins, Beard contends that the debate over the ratification of the Constitution took place basically between those with personal property against those with real property: "Substantially all of the merchants, money lenders, security holders, manufacturers, shippers, capitalists, and financiers and their professional associates are to be found on one side in support of the Constitution and . . . substantially all or the major portion of the opposition came from the non-slaveholding farmers and the debtors" (Beard 1961, p. 17). Critics have already shown the need to qualify this overdrawn distinction, for there are too many exceptions: wealthy merchants opposed and poor debtors favored the Constitution.

Let us instead show how our treatment of four of the leading Federalists contrasts with Beard's.

Benjamin Franklin. According to Beard (1961, p. 39), Franklin, "in the midst of his varied activities as printer, diplomat, statesman and philosopher, managed by thrift and investments to accumulate a considerable fortune for his day, about $150,000." Although Franklin indeed supported the Constitution, he "seems to have entertained a more hopeful view of democracy than any other member" at the convention, favoring a single-chambered legislature and resisting the attempt to place property qualifications on the electorate (Beard 1961, p. 197). Ironically, Franklin's views were considerably more democratic than, for example, Elbridge Gerry, who wound up opposing the Constitution, despite the fact that he was a wealthy merchant.

George Washington. Washington, says Beard (1961, p. 144) "was probably the richest man in the United States in his time, and his financial ability was not surpassed among his countrymen anywhere." Flexner (1984, pp. 368–370) confirms that "Washington's success as a land speculator was phenomenal" even while financial leaders including Robert Morris were going into debt. "Washington's financial success was based altogether on his own efforts. . . . Clearly, he had great business gifts."

Washington, of course, supported the Constitution, but Beard (1961, p. 214) concedes that his part in the proceedings "was almost negligible." When he did discuss the new system, it was in general terms, as for example in his Farewell Address when, with the assistance of Alexander Hamilton, he referred to the new government's being "the offspring of our own choice, uninfluenced and unawed, adopted upon full investigation, and mature deliberation" (cited in Beard 1961, p. 215).

Alexander Hamilton. Beard (1961, pp. 14, 107) observes that it is "highly improbable" that Hamilton ever held any considerable sum in public securities "for he was at no time a rich man" and "he died a poor man."

> Though he lived well, and had a large income apart from his paltry salary as Secretary, his earnings as an eminent lawyer may very well account for such sources of revenue as he may have enjoyed. Certainly, had he seen fit to employ his remarkable talents in private enterprise, he might have died one of the rich men of his day.

If Washington was perhaps an economic beneficiary of his Secretary of Treasury's economic policies, Hamilton himself apparently was not!

Beard (1961, pp. 100–106) nonetheless correctly points out that Hamilton explicitly emphasized the need to link support for the new political order with support for the prime commercial interests. Hamilton "had little part in the formation of the Constitution" (Beard 1961, p. 100), and as Beard (1961, p. 199) notes, what little influence Hamilton attempted to wield was rejected as too undemocratic, including proposals for a President and a Senate chosen for life. Thus both Franklin and Hamilton came from modest origins, both experienced a rapid rise, and both favored the Constitution. But the wealthy Franklin advanced one of the most democratic positions while Hamilton, "who was at no time a rich man,'' promoted one of the most anti-democratic positions. Both were rejected. (It is of note that it was another wealthy international cosmopolitan, Thomas Jefferson, who took noticeably more democratic—and localist—positions than Hamilton.)

Beard (1961, p. 100) nevertheless refers to Hamilton as ''the colossal genius of the new systems'' because ''it was his organizing ability that made it a real instrument bottomed on all the substantial interests of the time.'' It was, continues Beard (1961, p. 117), ''no mere abstract political science which dominated his principles of governments.'' Nevertheless, in view of his own modest economic standing, Beard (1961, p. 114) acknowledges, ''That he was swayed throughout the period of the formation of the Constitution by large policies of government—not by any of the personal interests so often ascribed to him—must therefore be admitted.''

James Madison. A descendant of one of the old landed families of Virginia, ''whose personal property was relatively small in amount,'' Madison himself

> seems to have relied upon the involvements of office and his father's generosity as a source of income. . . . His inclinations were all toward politics, for which he was prepared by long and profound research in history, law, and political economy. He was constantly in public life. (Beard 1961, p. 125)

Madison, Beard (1961, p. 205) notes, was ''the systematic philosopher of the Convention.'' Beard goes on to discuss his contribution in the context of the Federalist papers, notably number ten. Beard (1961, p. 125) does say that having devoted himself ''to political pursuits rather than commercial interests of any kind'' and having ''none of the public securities, Madison was able later to take a more disinterested view of the trading system proposed by Hamilton.'' By Beard's own economic ''interpretation,'' this observation should apply to the Constitution itself.

Now these four brief case studies of several of the Founding Fathers prove nothing, except perhaps that no single set of factors can account for their views. But, as noted, Beard himself rejects any simple determinism. I would only add that if economic factors cannot be ignored here, neither can cognitive ones, and Beard (1961, p. 251) would possibly agree for he observes that, ''generally speaking,'' both ''talent'' and ''professional abilities'' were ''on the side of the Constitutionalists.''

Beard himself (1961, p. 73) in effect introduces a significant cognitive variable in his economic interpretation. He rejects the position that the Framers necessarily benefitted personally from the Constitution. Rather, ''Did they represent distinct groups whose economic interests they understand and felt in concrete, definite form through their own personal experiences with identical property rights, or were they working merely under the guidance of abstract principles of political science?'' Beard here draws a somewhat spurious distinction between the ''concrete'' and ''abstract,'' seeming not to notice that large-scale economic interests are themselves an abstraction. Both Hamilton and Madison certainly were explicitly aware of the existence and importance of economic interests, but by

Beard's own account, neither had been personally involved in economic pursuits in any extensive way over the course of their own careers. Their analysis remained abstract.

Note too the levels of abstraction that are possible here: Will one identify with his specific trade or profession only (observe that this question can pose a jurisdictional problem in union organizing, for example)? Might one transcend economic class in part or altogether, as Franklin appears to do, and identify with the people more generally? Is this not what communist theorists from Marx and Engels through Lenin and Mao appear to do, rising above their bourgeois origins to identify with the lower classes, if not the entire population? And, of course, economic class can be qualified or even opposed by other considerations: ethnic, linguistic and religious, for example. Does one think in local or cosmopolitan terms? It may be quite appropriate to turn Beard on his head here. That is, perhaps Hamilton did not so much think in continental terms in order to further particular economic interests; rather he might have thought in terms of particular economic interests, as Benson (1960, p. 208) indeed argues, in order to further his vision of continental union. Benson thus cites Mitchell (1957, p. 422): "The exertion to promote the Constitution was not covert, for a class, but *pro patria*."

The Work of Orin Libby and Lee Benson: A Sectional and a Social Interpretation

According to Orin Libby (1894), a significant geographical as well as class division existed in the politics of the Constitution. The debtor areas of the frontier, the sparsely settled agrarian interior, were opposed to the creditor, mercantile urban centers near the coast.

Libby's sectional understanding goes back in time and influence to the seminal work of Frederick Jackson Turner, whose contemporary descendant, Jackson Turner Main (1961, p. 268), finds in this interpretation "much truth":

> There had most certainly been just such a division all during the colonial period, and it continued to be of great importance for many decades. There is an abundance of evidence to support this explanation; as a matter of fact, the contrast between seacoast and interior was in some cases even more marked than Libby believed.

Davis (1977, p. 152) also continues this sectional analysis, noting that even Federalists and Antifederalists were otherwise united by section, class and outlook. "Men like Elbridge Gerry, Richard Henry Lee, Alexander Hamilton, and James Madison had far more in common with each other than with rebellious debtor farmers in western Massachusetts."

Benson's "social interpretation" (1960, pp. 215–31) is similar. He distinguishes, in what he admits are oversimplified terms, between an agrarian and a commercial society in which "cosmopolitan commercialism" tended to be nationalistic and "localistic agrarians" parochial. Indeed, Main's localist/cosmopolitan distinction

(1973, p. 266) dovetails both Benson's social interpretations and Libby's sectional emphasis: Cosmopolitan strongholds were the well-populated urban coastal areas that were centers of commerce and power.

It remains, however, to place these perspectives within our populational cognitive context. Interestingly, Benson (1960, p. 216) in effect does so when he refers to his distinction as involving the "agrarian-minded" versus the "commercial-minded," that is, as not denoting "classes, socioeconomic status groups, or occupations" but rather "ways of thinking." "Thus, farmers with an intention or prospect of changing their occupation may have been commercial-minded, and men not engaged in farming may have been agrarian-minded." Main (1973, p. 46) also recognizes the danger of an overly typological approach when, for example, he describes "a set of attitudes" which distinguished two types of farmers:

> Because the great landholders enjoyed an income beyond subsistence, he could send his son to school or hire a tutor, buy books, subscribe to a newspaper, and travel. . . . These experiences widened his horizons. To be sure, some men of broad acres remained isolated in mind and body, but most became, by comparison with the rest of country folk, cosmopolitan and urbane. The same advantages made the large landowner more appreciative of the arts: he built his house according to a book on architecture, ordered some stylish furniture from England, hung up a portrait of his wife, heard a performance of chamber music (Washington himself played the flute) and attended a play. In many ways he felt himself more like the well-to-do townsman than the ordinary yeoman.

Even this distinction is too stark, but the major point remains valid: cognitive variables of an individual nature may override objective environmental categories. Thus if large plantation owners were in general cosmopolitan, some nonetheless remained "isolated in mind and body." But both locals and cosmopolitans could be further subdivided, and at one extreme of the local/cosmopolitan continuum, we might place James Madison, who in Beard's own words (cited in Benson 1960, p. 205) is described as follows: "Attached to his native state, he was nevertheless as much at home in New York or Philadelphia, as on his plantation in Virginia."

Note that the objective environment is often the result of self selecting forces at work. Thus a cosmopolitan by nature, despite lowly origins, will attempt to select a cosmopolitan environment. Both Benjamin Franklin and Alexander Hamilton, for example, migrated to congenial urban centers. Someone with a more local bent, such as a William Findley, may well, given talent and ambition, also migrate, but Findley initially established himself, not in Philadelphia, but in rural Waynesboro, Pennsylvania, before moving on to the western frontier in Westmoreland County. Many other frontiersmen were no doubt similarly self-selected. The larger point is that such self-selection may have a critical underlying biological dimension as well as important political implications.

A Self-Selection Interpretation

One of the most significant examples in American history of the self-selection process in operation is the 1787 Constitutional Convention in Philadelphia. Elkins and McKitrick (1978, p. 73) point to the "remarkably effective process of self-selection" at work for the convention: "There were leading state particularists such as Patrick Henry and Richard Henry Lee who were elected as delegates but refused to serve. The Anti-Federalists position indeed was hardly represented at all, and the few men who did represent it had surprisingly little to say. Yates and Lansing simply left before the convention was over." Rossiter (1966, p. 130) points out that if the Philadelphia delegates were not a representative mix, it was "in large part because of a natural process of selection (and self-selection)" had excluded the Antifederalists.

Now what, then, were the distinguishing features of the delegates who selected themselves. According to Elkins and McKitrick, there were three: (1) A continental outlook and background, (2) energy and ability and (3) youth.

A continental outlook. Elkins and McKitrick (1978, p. 66) contrast the continental outlook and experience of the leading Federalists with the careers and outlook of the Antifederalists, whose "constant preoccupation was state sovereignty rather than national efficiency."

Ability and energy. Elkins and McKitrick (1978, pp. 67, 68) point to the mobility of James Madison and Alexander Hamilton as "exemplars of nationalist energy": "In a number of outstanding cases energy, initiative, talent, and ambition had combined with a conception of affairs which had grown immense in scope and promise by the close of the Revolution."

Youth. Elkins and McKitrick stress the relative youth of the leading Federalists as opposed to the Antifederalists: "A significant proportion of relative newcomers, with prospects initially modest, happened to have their careers opened up at a particular time and in such a way that their very public personalities came to be staked upon the national quality of the experiences which had formed them." The leading Antifederalists, on the average ten to twelve years older (Elkins and McKitrick 1978, p. 66), saw their careers significantly begin in a prerevolutionary era that did not encourage continental thinking.

No doubt territorial and temporal conditions influence human action, but again, individual cognitive responses must be kept in mind. Elkins and McKitrick rely on Jensen (1958) for a listing of leading Federalists and Antifederalists. Jensen (1958, pp. 424–25) bases his listings on the time period of the Declaration of Independence in 1776 to the convention of 1787. Thus, he somewhat arbitrarily excludes Benjamin Franklin, who, as we have seen, was thinking in continental terms beginning with the French and Indian wars, some twenty years before the American Revolution and a good generation before the Philadelphia convention.

A second problem with the listing is that the Antifederalists named are necessarily the most prominent, but not necessarily the most typical. Here is Jensen's list (1958, p. 424): "Samuel Adams, Patrick Henry, Richard Henry Lee, George Clinton,

James Warren, Samuel Bryan, George Bryan, Elbridge Gerry, George Mason and a host of less well known but no less important men in each of the states.'' Notice the last phrase—''and a host of less well known but no less important men in each of the states.'' Following another lengthy listing of Federalist leaders, Jensen (1958, p. 425) ends more simply: ''and many lesser men.'' In other words, as Wood (1969, p. 426) maintains, the more typical Antifederalists were also the result of the American Revolution, which was marked by ''the sudden appearance of new men everywhere in politics and business.'' The emigration of thousands of Tories, the democratization of politics, and the breakup of older mercantile arrangements provided ''new opportunities for hitherto unknown but ambitious persons to find new places for themselves'': ''For every brilliant provincial lawyer ready to challenge the supremacy of the imperial clique in the colonial metropolis, there were dozens of lesser men, not so brilliant but equally desirous of securing a local magistracy, a captaincy of the militia, some place, however small of honor and distinction.'' One of these ''uneducated parvenus,'' according to Wood (1969, p. 480), was ''the ex-weaver William Findley.'' Thus, the period following the Revolution was not a broadening experience for those who, however young, were of a more localistic bent. James Madison, on the other hand, was already well on a nationalistic road by the time of his graduation from Princeton, five years before the Declaration of Independence.

If we grant that the leading Federalists were by and large young, able, energetic, and nationalistic and that they self-selected themselves in Philadelphia, what follows next? Here Elkins and McKitrick (1978, p. 73) are right in their analysis, for they see correctly the synergistic effects of the self-selection process: It produced ''a growing esprit de corps'' that made the delegates ''more and more committed to the product of their joint efforts'' and that ''was in all likelihood the key mechanism, more important than any other in explaining not only the peculiar genius of the main compromises but also the general fitness of the document as a whole.'' Rossiter (1966, pp. 130, 131) concurs, referring to ''a pervasive, developed esprit de corps from the beginning'' as a ''happy reunion of old friends and comrades'' took place: ''They were, indeed, a continental elite, the nearest thing to an 'establishment' that could have existed in those days of poor communications, limited horizons, and divided loyalties.''

Where Elkins and McKitrick (1978, p. 72) err is in calling the convention ''a really classic study in the sociology of intellect.'' A more accurate reference would be to the ''biosociology of intellect.'' As discussed previously, an important biological dimension underlies groups formed by self-selecting mechanisms. In the case of the Constitutional Convention, men of generally cosmopolitan space-time horizons reinforced each other's natural propensities. They in part formed the environment that in turn motivated them.

A Madisonian Understanding

Madison seemed to be of two minds concerning the factors that led to the formulation and adoption of the Constitution. In advocating the adoption of the Constitution in the *Federalist Papers*, especially No. 10, he seemingly took a hard realistic, deterministic view of the causes of human action in general. Yet, especially in speaking directly of the Constitution, as well as of the process envisioned for the selection of national leaders, Madison saw a role for rational disinterested choice. We will briefly examine both sides of Madison's thought.

In *Federalist No. 10*, Madison sets forth his classic position: "But the most common and durable source of factions has been the various and unequal distribution of property. Those who hold and those who are without property have ever formed distinct interests in society." This statement is an apparent economic determinism that anticipates Marx and Beard.

We must, however, qualify Madison's position in three ways. First, note that Madison refers to the presence and importance of the "various" as well as "unequal" distribution of property. In other words, as Diamond et al. (1972, pp. 92–94) observe, for Madison, people within the same general economic class may nonetheless be torn and influenced by conflicting interests.

Second, as Brown (1956) notes, Beard neglects to refer to those passages in *Federalist No. 10*, in which noneconomic influences are explicitly mentioned: "A zeal for different opinions concerning religion, concerning government, and many other points." Clearly Madison is no simple economic determinist (nor, for that matter, is Beard).

Third, in the same *Federalist No. 10*, Madison later refers to at least the possibility of "enlightened statesmanship." Granted Madison says that "enlightened statesman will not always be at the helm" to deal impartially—or to try to so deal—with competing factions. Madison then outlines a political system that will moderate the "mischiefs" of faction independently of the presence of enlightened statesmanship. Yet one way in which faction is moderated in the new constitutional order is by a system of representative government whereby the outcome of elections tend to result in the choice of "men who possess the most attractive merit." The critical point here is not that Madison is necessarily right in his claim, but that he believes that such a claim may be validly made, that is, that enlightened leadership can exist and can make a difference. As Benson (1960) notes, Beard also acknowledges such a possibility—precisely with reference to the Father Founders in his own preface for his book, *Enduring Federalist*.

This last point suggests the second side of Madison: the possibility, at least, of relatively objective rational political leadership. Indeed, Madison (1987, pp. 18, 19) saw the "architects of the Constitution" in such a light. He wrote that "there never was an assembly of men, charged with a great arduous trust, who were more pure in their motives" than "were the members of the Federal Convention of 1787" (cf. Madison, *Federalist No. 37* and Meyers 1981).

It was this aura of aristocratic virtue that the Antifederalists wished to dispel. "Instead of seeing enlightened patriots," Wood (1987, 93) writes, "simply making a Constitution to promote the national good, they saw groups of interested men trying to hoist an aristocracy onto Republican America." Wood (1987, p. 93) concludes that "they, not the Federalists, were the real modern men." Their understanding of American society was "far more hardheaded, realistic, and prophetic than even James Madison's."

It is beyond the scope of this volume to resolve this question concerning the possibility of disinterested statesmanship. But our formulation of a local-cosmopolitan continuum may be suggestive in this respect. The more extensive one's space-time horizon, the easier it becomes to transcend immediate environmental influences, just as such a possibility is relatively realized for most people in the process of maturation. As one approaches the status of international cosmopolitan, the impact of ideas becomes more critical. Now the virtue of the ideas will depend on the virtue of the individuals who frame them and not just on their intellect. The Founders explicitly recognized the necessity for both ability and character in political leadership, and if their constitutional arrangement did not always ensure its presence, their own efforts at least reflected those characteristics. The Founders were, in any event, cosmopolitans and would not have succeeded in nor even attempted a nationalist revision of the Articles of Confederation had they not been.

The biological basis for differing local and cosmopolitan space-time horizons was in effect recognized by Madison, for he explicitly refers to and emphasizes in *Federalist No. 10*, the "diversity in the faculties of men." Such a diversity may indeed cause class division, but it also opens up the possibility for the emergence of able enlightened leadership.

And so the Madisonian understanding anticipates our own.

NOTE

1. Cf. Mitchell's comparison of Washington and Hamilton (1957, p. 232): "Washington was less intellectual, was equal in reasoning power, superior in judgment. He was not so quick, but was surer. What he lacked in detail of information he more than made up in sum. What glittered in Hamilton was gold, but had its modicum of alloy. Washington's metal was purer."

Part III

The Neuropolitics of Alienation

Chapter 6

Gene-Environment Interaction: The Neuropolitics of Alienation

> Perhaps the crucial underlying fact of Sullivan's affective life was the high level of his anxiety, indications of which pervaded his history and his behavior and test records at the Clinic . . . the theme of being 'trapped' ran through several episodes in which his enterprises were at a nadir, and recurred in his Free Association Hour.
>
> —Smith, Bruner, and White on "Sullivan," *Opinions and Personality*

> His current life was the source of great satisfaction. About his legal training and his work at the office he wrote as follows: "The school is wonderful. For the first time in my life I find myself looking forward to classes. I believe I am at or near the head of my class. The case method was made for me. As though that weren't enough I am equally enthusiastic about my work.
>
> —Smith, Bruner, and White on "Chatwell," *Opinions and Personality*

> Of the future of his job, he says: "There isn't any future in that job. I mean as far as I'm concerned, this is about as high as I could go . . . if I was feeling well and didn't have this ulcer, I'd probably like the job, see. Those things (worries about supplies and the boss's failure to help him) wouldn't bother me so much, see? I mean those things are—why, they just become enormous problems, you know what I mean?"
>
> This man is not distant from his work; he is overwhelmed by it.
>
> —Lane on "Rapuano," *Political Ideology*

According to Darlington, each individual "seeks and selects the place, the environment that fits" him or her "best" (1969, p. 269). The fatal flaw in Darlington's hereditarian formulation lies in its failure to recognize that individuals may be denied access to those environments most in keeping with their genotypic endowment. Nevertheless, with this vital qualification in mind, Darlington's emphasis on self-selection may properly serve as an antidote to a prevailing environmentalism, one that tends to assume a given phenotype or social characteristic—for example, a status, role, or attitude—to be an exclusive outcome of external social forces.

In this chapter, we propose an interactional model in which individuals may actively seek suitable environments. If individuals attempt to select the environments that fit them best, not every attempt will prove successful. Many individuals will find themselves stifled by oppressive environments. Still others, through accident of birth or through pressing persistently onward in "Peter Principle" fashion, may face circumstances beyond their innate capacities for coping adequately. Thus at any one time, an individual may feel trapped, satisfied, or overwhelmed by the environment.

These perceptions are phenotypic reflections of genotype/environment interactions, or more specifically, of three basic possible combinations of genetic capacity and environmental opportunity:

1. less external opportunity than innate ability (entrapment)
2. more opportunity (opportunity as imposition) than ability (overwhelmingness)
3. opportunity and ability in balance (satisfaction)

This simple interactional analysis discloses two qualitatively distinct forms of personal alienation: entrapment and overwhelmingness. The environmentalist emphasis, on the other hand, permits only the classical form of alienation, that of entrapment. Environmentalists cannot admit that circumstances may be overwhelming, at least, insofar as they tacitly adhere to a premise that is egalitarian and utopian, one that endows every individual equally with seemingly unlimited potential. Hereditarians, by contrast, can more easily view an individual as the prisoner of a limited genetic capacity than of an alienating environment.

In addition to the importance of each of the three basic genotype/environment interactions, we will indicate that a significant neurobiological dimension exists, since the human capacity to envision alternative environments is likely an evolved neurophysiologically based faculty. In particular, it can probably be related to the development of the frontal cortex. As Hynd and Willis (1985, p. 129) observe: "The advanced development of the frontal lobe is truly what distinguishes humans from other animals. In general, it can be said that the frontal lobes comprise about 40% of the total area of the brain in humans." The development of such a large brain area probably holds some significance in accounting for human action. According to Bloom, Lazerson, and Hofstadter (1985, p. 209), "Accounts of people with damaged frontal lobes . . . do attest to the critical role of these lobes in judging and planning . . . they become irresponsible, unable to act appropriately or to carry out any consistent life plan." When we later refer to Goldstein's (1963) classic study of brain damaged patients, a specific instance of such severe impairment in human functions will be provided. Since such impairment was not genetic in origin (it was the result of wartime injury) but nonetheless exists in organic form, we will refer more usefully to an organism/environment rather than genotype/environment interaction, with the organism representing the central nervous system of an individual.

Finally, we shall argue that each of the three interactions implies a distinctive set of political responses on the part of the individual. For example, the entrapped will tend to adopt views of a politically radical coloration. The overwhelmed will tend toward a reactionary orientation, while a moderate-to-conservative stance appropriately reflects the position of the satisfied. These assertions are subject to qualification on both a micro and macro level and their speculative nature must necessarily serve to raise questions rather than resolve them.

ENVISIONING ALTERNATIVE ENVIRONMENTS

Dobzhansky (1970, p. 25) observes that "really successful species are masters of a variety of environments" and that the human species is "perhaps most adept at this strategy." From instinctive preference to conscious choice, from rigid specialization to flexible mastery, from amoeba to humans—a great deal of ground is covered by the rubric of self-selection. But since we are directly concerned with human beings—who can, as Dobzhansky noted, "choose, control, and to some extent create environments"—we are also primarily concerned with the capacity of humans to exert conscious control over the environment.

Human intelligence involves the capacity to envision, seek, select, and create alternative environments. (cf. Itzkoff 1991). The capacities for intelligent choice, for self-selection, and for evolutionary survival should therefore all be interrelated. All societies, Dunn and Dobzhansky (1952, p. 69) write, "primitive as well as advanced, by their existence and survival place a premium of brain power, on the ability of the individual to learn from experience and to modify his behavior accordingly." As Simpson (1964, p. 276) asserts: "Probably we could all agree that the most distinctive human characteristic now and the one most important for our future is intelligence."

The presence of genetic diversity within populations implies a wide range of capacities for self-selection and a similar range in capacities for conceptualizing environments. In the remainder of this section, four gradations of cognitive ability will be proposed to correspond to four capacities for self-selection that Darlington has developed.

These gradations are conceptually based on the abstract-concrete distinction first employed by Goldstein and Scheerer (1941, pp. 3–4). In their work dealing with brain-damaged patients, they saw their subjects as characterized by the "concrete attitude": They "are confined to the immediate apprehension of the given thing or situation in its particular unique sense. . . . [Their] thinking and acting is directed by the immediate claims which one particular aspect of the object of the outer world makes.''

The normal individual, on the other hand, possesses the ''abstract attitude.'' According to Harvey, Hart, and Schroeder (1961, p. 4), this individual operates more ''in terms of multiple alternatives (within a more complex and dimensionalized space) rather than in terms of bifurcated black-white categories.''

The ''objective'' environment assumes reality for an individual on the basis of ability—large or limited—to conceive it. According to Harvey, Hart, and Schroeder (1961, pp. 24–25):

> Concreteness-abstractness is assumed to be an attribute of the concepts or patterning of stimuli and not of the stimuli themselves. . . . The greater one's abstractness (1) the greater is his ability to transcend immediately and move into the temporally and spatially remote, (2) the more capable he is of abstracting relationships from objects of his experience and of organizing them in terms of their interrelatedness.

Two key aspects of the concrete-abstract dichotomy require qualification. Goldstein (1963, ch. IX) speaks of all normal (i.e., non-brain-damaged) individuals as sharing the abstract attitude but with individual ''preferences'' for the more concrete or the more abstract. Hence, the use of the term *attitude*. There is thus, rather ironically, an egalitarian and abiological bias in his work.

This bias, however, is not warranted. In order to examine brain-damaged patients, Goldstein and Scheerer devised a number of sorting tests. These tests require the subject to take objects of diverse shapes and colors and to sort them along as many dimensions as he or she is capable of conceptualizing. In this way, a measure of abstract capacity is obtained. As Anastasi (1968, p. 310) points out, ''The Goldstein-Scheerer tests have been found to correlate highly with scores on standard verbal intelligence tests.'' This should not be surprising, since both tests put a premium on the ability to differentiate, to analyze, and to generalize. Like standard verbal intelligence tests, one's level of performance on the sorting tests should reflect a largely inherited capacity (cf. Vandenberg and Vogler 1985).

Concept of Intelligence

The concept of intelligence remains a highly elusive and controversial one. Before relating it to our concept of space-time horizons, with its dimension of abstract ability involving varying capacities to envision alternative environments, we first must distinguish between three approaches to human intelligence: ethological, cognitive, and psychometric.

Ethological

''Cognitive abilities are viewed here as evolved adaptations that aid animals in their interactions with complex, variable, and hazardous features of their environments'' (Charlesworth 1983, p. 238). From this perspective, intelligence constitutes "the way one knows the world and uses that knowledge when adapting to changing situations." Hence "[b]iological intelligence may be nothing more (or less) than the capacity to construct a perceptual world. For man this is the real world of which he is conscious" (Jerison 1973, pp. 16, 17; cf. Losco 1982).

According to this evolutionary analysis, "all animals with sufficiently large brains . . . create their various real worlds" and these "worlds will differ from one another just as the brains that create them differ" (Jerison 1977, p. 54; cf. Hynd and Willis 1985). Thus, for Jerison, the "concept of space, time and objects" are viewed as creations of evolving nervous systems.

In human evolution, Etkin (1985, p. 499) stresses the role of "projective consciousness," which he defines as "the ability to project consciousness beyond the immediate present, as apprehended by our sensory system, to the past (memory) and to the future (anticipation)." It is, for Etkin (1985, p. 500), "the critical factor accounting for the development of culture-capabilities in man." Clearly, Etkin's "projective intelligence" constitutes the temporal dimension in our space-time horizon, which Etkin (pers. comm.) confirms.

The space-time horizon, therefore, can be viewed as an evolutionary and neurobiologically based human adaptation to the environment and as such would constitute a critical component of human intelligence (cf. Stenhouse on "the evolutionary intelligence and human futures" [1973, pp. 177–85]). Our major qualification of this formulation would be that, in its most active and abstract form, the space-time horizon permits much more than an adaptation to the environment in a passive sense, but rather the major alteration of existing environments or even the creation of entirely new ones.

Cognitive

Cognitive science "is the study of the human mind as subjected to the scrutiny of the allied disciplines of psychology, computer sciences, philosophy, and linguistics. . . . Cognitive scientists are particularly interested in human intelligence, because intelligence represents, in some sense, the epitome of human functioning—that which makes us distinctly human" (Sternberg 1986, p. 137).

This statement by Sternberg, perhaps the leading cognitive scientist in the area of human intelligence, points to the differences and similarity between the cognitive and ethological approaches. Both are similar in acknowledging the central role that intelligence plays in human life. Both see it as playing adaptive or functional roles.

The mode of analysis, however, differs. The ethologist stresses the observation of life in naturalistic circumstances. The cognitive scientist, eager to utilize the latest technological tools especially in the form of the computer, tends to confine his or her analysis to laboratory conditions—although Sternberg (1985, p. 109) does concede "a pressing need to investigate the components of intelligence as they operate in everyday life as well as in fairly artificial laboratory settings."

In Sternberg's analysis of intelligence (1986, p. 140), a distinction is drawn between "the metacognitive processes that drive the cognitive ones." The cognitive processes involve the abilities required to solve varying problems along a wide range of areas, including linguistic, spatial, and mathematical. The metacognitive processes, on the other hand, include those involved in "recognizing that a problem exists, defining the problem, constructing a strategy for the solution of the problem, selecting a mental representation on which the processes and strategy will operate (e.g., a propositional representation or spatial representation), allocating both internal and external resources to the problem, and monitoring one's solution of the problem." Sternberg (1986, p. 142) draws a similar distinction between "global" and "local planning" as strategies in problem solving. Local planning takes place at "a specific level, whether for a single test item or from one stage of one's career"; global planning, on the other hand, occurs "at the general level, whether for an entire set of test items or for one's career."

Clearly, it is the metacognitive processes with a special emphasis on its global planning component that corresponds most closely to the space-time horizon, as well as to the ethological intelligence as discussed above. Of course, the space-time horizon also relates to more specific abilities, involving both spatial and temporal factors. As already stated above, the spatial component of the horizon relates closely to Gardner's general definition of spatial ability (1983). It would not, however, necessarily relate closely to any one specific spatial ability, as measured by any one specific test. Indeed, it is in this more operational, rather than broader, conceptional sense that "the construct of spatial ability remains somewhat ill-defined to the present day" (Sternberg 1985, p. 103). Intelligence tests share this difficulty insofar as many of them incorporate items measuring spatial ability.

Psychometric

According to Sternberg (1985, p. 59), "Cognitive theories elaborate on rather than replace psychometric ones. . . . Psychometric theories deal with intelligence primarily in its structural aspects; cognitive theories deal with it primarily in its processing aspects."

The psychometric approach is the older one, and although it is the more descriptive, it is also the more controversial. However, it may not be any more problematic than either of the cognitive or ethological approaches, but because of its concentration on the measurement as well as description of intelligence, it is the one approach that has directly measured differences in intelligence by group as well as by individual and related those to a possible genetic basis (see Kaplan 1985).

A central egalitarian response to such findings is to deny that such a thing as general intelligence exists (e.g., Gould 1981). Indeed, such a response has extended to the cognitive sciences, as in Gardner's emphasis (1983) on distinct abilities. Such an emphasis appears especially justified in the light of a tendency in the past for cognitive scientists "to make comparisons between groups, and to ignore quite glaring differences between the performance of individual subjects." But now there is an "increasing recognition of the existence of wide-ranging individual differences in cognitive differences" (Cohen 1983, p. 6).

The replacement of a typological perspective by a populational one is also urged by Sternberg (1985, p. 109, emphasis in the original): "It has been customary in much psychometric and cognitive research to pursue a *nomothetic* approach to understanding individual differences. In this approach, it is assumed that all individuals have essentially the same abilities, whether these abilities are measured by factors or components. Thus, the factors and components of mental abilities could be the same across subjects, with only their values differing in quantitative fashion. It now appears, however, that individuals differ qualitatively as well as quantitatively and that, as a result, a more *idiographic* approach is needed for the study of human abilities.''

Consequently, not only may general intelligence not exist, but specific mental abilities may be reflected in unique ways in each individual. Although a populational emphasis is long overdue, it should not be extended to the other extreme by denying any similarities in human cognition. With respect to the long-standing controversy over general versus specific intelligence, for example, Sternberg (1985, p. 110) notes the current emphasis in cognitive psychology on ''domain-specific abilities,'' but nonetheless states that ''the appearance of general and major group factors in intelligence tests argues strongly for the generality of at least some abilities.'' Sternberg suggests in particular that ''those represented by metacomponents of executive processes'' are ''quite general.'' It is the metacognitive capacity to envision alternative environments that would fit this description of a ''quite general'' ability.

Note how the space-time horizon could be construed in an entirely populational perspective. It could, following Gardner (1983), be wholly separated into its spatial and temporal dimensions. It is conceivable that, at least for some individuals, an extremely high spatial ability may be combined with an extremely low temporal capacity. Gardner's own examples of idiot-savants provide a graphic and valid illustration. Yet the opposite point can be made as well: very few of us are idiot-savants, in which abilities are so disparately related.

Self-Selection and Intelligence

The position required is one that avoids either a wholly typological or populational position. Hence the concrete-abstract dichotomy described thus far is too stark. Indeed, if each human genotype is unique, so is each individual intelligence as a genotypic capacity. Therefore, intelligence is not wholly a unitary trait, but to the extent that it is, the most plausible assumption would hold that it is normally distributed for a large population. To be sure, this assumption cannot be directly tested and demonstrated, but it is more fully in accord with contemporary scientific knowledge than the environmentalist position still in vogue—which on the surface professes an abiding agnosticism on the question but proceeds tacitly to assume an egalitarian distribution of natural talents that is distorted only by inequities in socioeconomic opportunity.

The classification of abilities about to be introduced avoids, on the one hand, a stark dichotomy and, on the other, a continuous distribution that, lacking conceptual distinctions, is without any convenient handle. Yet any typological classification is as destructive of reality as it is congenial and, hence, essential to scientific inquiry. The four categories to be presented here do not escape this stricture. They too, in Wordsworth's apt phrase, "murder to dissect."

At least, however, they have the virtue of corresponding with Darlington's (equally arbitrary and useful) conceptual scheme. With respect to their capacity for self-selection, Darlington (1969, p. 271) places individuals within the following four orders:

1. accepts the environment in which they find themselves.
2. seeks an environment that suits them.
3. changes their environment to suit themselves better.
4. changes the environment of others for their own and even for succeeding generations, for their own species, and even for the whole world.

Now, individuals fit within our four gradations of abstract capacity as follows:

1. The concrete, who lack the capacity consciously to conceive the environment.
2. The narrowly abstract, whose capacity to conceptualize the environment is fundamentally limited to their immediate time and place.
3. The functionally abstract, who possess the capacity to conceptualize an environment alternative to their existing one.
4. The highly abstract, who possess the capacity to conceptualize multiple alternative environments.

The intended correspondence between these two sets of categories is obvious and needs only to be explicit:

The concrete. Unable to conceptualize any environment, present or future, the concrete person is restricted by default to whatever environment happens to envelop him or her at the moment. He or she therefore falls within Darlington's first order: those who simply accept the environment as it is.

Goldstein's brain-damaged subjects illustrate this category, as do severely mentally retarded individuals. As Sternberg (1986, p. 140) observes, "A major feature separating the retarded from the normal individual is the retardate's need to be instructed explicitly and completely as to the nature of the particular task he is being asked to solve and the way it should be performed." Ross et al. (1985, p. 82), in a report of their forty-year follow-up study of 160 mentally retarded San Francisco residents whose mean childhood IQ was 67, describe them as being "generally unambitious, passive, and lacking in self-direction."

The narrowly abstract. For this individual, the here and the now are so immediately in conceptual focus as to exclude more distant spatial and temporal concerns. Consequently, like a camera in a fixed position for shooting, the person in this category will take advantage of any suitable change within the "narrow confines" of his vision; but he will not, because he cannot, consciously change his situation dramatically for better advantage—although external forces of dire proportions might impose reluctant dislocations. This individual fits Darlington's second order of individuals: those who seek an environment that suits them—the most numerous, according to Darlington, of his four orders.

The functionally abstract. Able to conceive an alternative environment, this individual may find the occasion and the incentive to try to change his or her existing one. But if the individual consciously pursues some aim beyond his or her immediate world in time and space, he or she will likely pursue it single-mindedly, for while the individual may on the surface show adeptness at dealing with a number of abstractions at once, especially if a prior goal is advanced by doing so, only one dominant concern at a time will have salience for him. The name for this category, accordingly, is influenced by Mannheim's (1940, pp. 51–54) concept of "functional rationality," which refers to the fact that a "series of actions is organized in such a way that it leads to a previously defined goal, every element in this series of actions receiving a functional position and role." Such individuals correspond to Darlington's third order—those who change their environment to suit them better—and they are presumably responsible for most of the social and spatial mobility that occurs within society.

The highly abstract. This individual possesses the capacity simultaneously to conceive and perhaps even to coordinate and pursue a number of alternative plans or visions. Here, exceptional capacity may often take highly individualistic form, but two general subcategories can be distinguished: ideas and action. Those individuals in the first subgroup can be called the "intellectually abstract" and are concerned with abstractions qua abstractions. Those in the second subgroup, or the "pragmatically abstract," may feel at home in the world of ideas, but are apt to use abstruse ideas (and their authors) in order to advance their own more tangible objectives. Like the functionally abstract, they are capable of singleminded pursuit, but beyond the merely functional, they may display a genuine interest in ideas, as well as the capacity to plan and coordinate a variety of projects at a time—a prime requisite for any top business executive, military commander, or public administrator.

The highly abstract subsume Darlington's fourth order—those who can change their environment both for themselves and for others—and they include, in Darlington's (1969, p. 272) words, "the Heroes of former ages, the Great Men of our own." The social significance of this grouping is, therefore, in inverse reaction to its numbers. From a larger, evolutionary perspective, "J.B.S. Haldane and others have," according to Mayr (1967, p. 836), "pointed out that the human achievements—those things that distinguish men from the animals and are responsible for man's civilization, art, literature, and science—were achieved by less than 1 percent of the human population, by those in the upper tail of the curve of human variation in inventiveness, imagination, perseverance, and ability to think clearly."

One caveat bears repeating before ending this section: A capacity does not imply its realization. It may serve as a necessary condition for that realization, but never

never as a sufficient one. Talent is one gift of life whose birth may precede its abortion.

ORGANISM-ENVIRONMENT INTERACTION: THE TRAPPED, THE SATISFIED, AND THE OVERWHELMED

If individuals vary in the capacities they possess, environments vary in the demands they impose. This section will show how the juxtaposition of a wide range of varying individual capacities, together with a similar range of environmental demands, yield importantly different sets of organism-environment interactions. In particular, four gradations of environmental demands will be juxtaposed against the four gradations of abstract ability just presented.

Our classification of environments is not necessarily limited to an occupational milieu, nor does it evaluate specific environmental settings in advance. But it does share in common with the Barr scale of occupations (cf. Sorokin 1959, ch. II) the principle that environments can be evaluated by the cognitive demands they impose on the individual. For our purposes, environments will be divided into four categories according to the following degree of abstract ability they demand:

1. concrete (work of a simple unskilled nature)
2. narrowly abstract (most blue collar and clerical labor, especially routine and semiskilled)
3. functionally abstract (most white-collar jobs, often with suggestive general designations, such as bureaucrat, functionary, organization man)
4. highly abstract (top professional and managerial positions)

Two points require clarification before proceeding. First, the examples given to illustrate these categories were, as in the Barr scale, occupational descriptions (note the centrality of job satisfaction in "overall life satisfaction" in Argyle 1987, p. 42), but the environment refers, beyond the world of work, to any salient features of one's surroundings. Ultimately, the environment must be defined for each individual, with great weight placed on each individual's own utilization of it. Even a prison environment, for example, cannot be judged categorically. It will vary by institution, by inmate, and by individual circumstance, especially insofar as some room remains within it for self-selection. Thus solitary confinement will inhibit almost any creative effort, but otherwise, self-imposed demands may succeed, at least in part, in transforming a penitentiary into a place of learning and creativity.[1]

Second, an emphasis on the single dimension of abstract ability clearly ignores other significant variables involving aptitude and temperament. Such limitations are undeniable. Other researchers, however, will have to compensate for these admitted inadequacies. The purposes of convenience and congeniality require simplification.

No ideal environment exists for each genotype but rather a certain range of suitable environments, for as Dobzhansky (1970, ch. 9) rightly insists, in the absence of a rigid genetic determinism, a degree of plasticity and educability does obtain. In addition, this ideal range of environments will still exist only in relation to the development of the organism, that is, of its physical phenotype from birth through death; consequently, these environments will change from one stage of development to another. Given for each individual in society his or her own ideal range of environments, which is in turn ever changing, the task of specification for larger analytical purposes becomes overpowering.

Let there at least be a starting point. A social analysis confined to the factor of abstract ability is clearly limited, but it does isolate a critical factor in the self-selection process—and one that has too often been neglected within the social sciences during the past several decades.

Environmental demands may now be related to individual capacities. External demands may fail to match innate ability (producing entrapment), more or less match it (satisfaction), or exceed it (overwhelmingness). These organism-environment relationships may be set forth more systematically by employing four categories of abstract ability for both the individual genotype and the environmental setting as shown in Table 6.1. The three basic relationships outlined in the figure are examined below.

Entrapment. The possibility that self-selection may fail is clearly left open in this analysis. We need only to reexamine Table 6.1 to observe the number of positions in which individuals possess more ability than the environment allows them to express. Thus it is an open empirical question as to the actual juxtaposition of individuals and environments in society.

No biological determinism, then, is formulated here. Nor the obverse, for genotypic capacity denied expression does not thereby cease to exist. Instead, it becomes

Table 6.1
Organism-Environment Interaction for Abstract Ability

Organism	Environment			
	i Concrete	ii Narrowly Abstract	iii Functionally Abstract	iv Highly Abstract
1. Concrete	S	O	O	O
2. Narrowly Abstract	T	S	O	O
3. Functionally Abstract	T	T	S	O
4. Highly Abstract	T	T	T	S

Key: S = Satisfied
O = Overwhelmed
T = Trapped

frustrated and forms the basis for the classical understanding of alienation: that a person is in possession of a true or inner self which has been stifled. For the intellectually abstract individual, in particular, the possibility of such entrapment is ever present. Since no society sustains more than a relatively small number of intellectually abstract positions, an accident of birth or of circumstance may well deny such an individual an optimal milieu, talents and aspirations notwithstanding. But for the less gifted, especially the concrete person, entrapment is not likely to occur. Exposure to an inhumane environment is, of course, all too possible, but on a cognitive level, almost all social environments presuppose at the least a narrowly abstract capacity.

Satisfaction. Argyle (1985, p. 48) notes that "generally, satisfaction is higher when there is a good fit between person and job," and that "depression results if the job is too complex or not complex enough." If all individual abilities and environmental demands in society were more or less in balance (position 1, i; 2, ii; 3, iii; 4, iv; in Figure 6.1), the outcome would constitute a definition of meritocracy and the basis for a stable political order.

Stability, however, does not imply sameness. Our analysis, typological though it may be, nonetheless warns against over-typologizing: Even if everyone in society were "satisfied," they would still display different interests and attitudes, varying from the concrete (1, i) to the highly abstract (4, iv). A similar observation applies with equal relevance to the entrapped and the overwhelmed. (Cf. the "flow channel" in Csikszentmihalyi 1990, p. 74.)

Overwhelmingness. An individual may be neither stifled nor satisfied, but be overwhelmed with responsibilities. Far from being bored or oppressed, an individual's energies and abilities may be stretched to the breaking point, producing feelings of fatigue, frustration, and powerlessness.

Even the highly competent may prove inadequate to the most pressing of social responsibilities, especially as many of these grow ever more awesome. It is suggestive that the American best-selling work of nonfiction throughout 1969 was *The Peter Principle* (Peter and Hull 1969), with its central proposition that "In a Hierarchy Every Employee Tends to Rise to His Level of Incompetence." The rise of meritocracy and the Peter Principle may go hand in hand (cf. Neff 1985, ch. 13; Masters 1986).

More generally, societies may change over time, with a reordering of the positions of individuals in relation to their environments. Concomitantly, individuals, over their own lifetimes, may change their circumstances appreciably. Thus, a bright young subordinate might initially feel some entrapment, subsequently advance in standing and experience satisfaction, and ultimately prove the Peter Principle by confronting tasks more imposing than his or her ability. (Or then again, at the height of success, he or she may suddenly fail.) It is, therefore, an open *and* continuing question as to the position a person might occupy over a period of time.

The Politics of the Trapped, the Satisfied, and the Overwhelmed

In this final section, individual case studies shall be presented to illustrate the previous analysis, and a number of political implications will be explored on individual and societal levels. Two studies that deal intensively with a small number of Americans and their political attitudes during the decade following World War II are useful to our analysis: Smith, Bruner, and White's *Opinions and Personality* (1964) and Lane's *Political Ideology* (1962). Table 6.2 reintroduces the organism-environment interactions for abstract abilities with the categories filled by names of subjects examined in the two studies.

We will group these names together under the general categories of entrapment, satisfaction, and overwhelmingness and examine each category in turn, beginning with the most numerous category.

The Satisfied

If individuals succeed in finding suitable environments for themselves, it follows that they will be satisfied with the existing order of things. They will feel "allegiant," using Lane's phrase (1962, ch. 10), to the prevailing political order. But as generally true as this analysis might be, it quickly becomes qualified when individual cases are considered.

Dempsey. Lane (1962, pp. 4, 10) introduces his study of fifteen, white lower middle class and working class voters in New Haven, Connecticut, by telling us that he is concerned with "not the unique but the common." Yet at the outset,

Table 6.2
Organism-Environment Interaction for Abstract Ability, with Individual Cases for Illustration

Organism	Environment i	Environment ii	Environment iii	Environment iv
	Concrete	Narrowly Abstract	Functionally Abstract	Highly Abstract
1. Concrete				
2. Narrowly Abstract	Dempsey*	Hodder	Rapuano*	
3. Functionally Abstract		Daniel	Lanlin	Clark
4. Highly Abstract			Sullivan	Chatwell

Sources: *R. Lane, *Political Ideology*, Glencoe, Ill.: The Free Press, 1962; all other cases are from M. Smith, J. Bruner, and R. White, *Opinions and Personality*, New York: Wiley, 1964.

he also points out that his conversations with the fifteen men were recorded and that they varied ''in size from one that is 154 pages 'written' by Dempsey, a drill-press operator of limited education and intelligence and small political interest, to one 322 pages long, 'written' by Flynn, a railroad supply clerk with a high school education, unusual intelligence, and a reflective cast of mind.'' Lane's observation serves to alert us to the inevitable presence of individual differences for any grouping, whether his own sample of the ''American Common Man'' or our present assemblage of the satisfied.

Lane's observation also suggests the classification of Dempsey as a narrowly abstract individual within a narrowly abstract environment: ''With a regular job, two girls, a boy, and a solicitous wife, yet with a limited intelligence, he has made a modest success of his life'' (1962, p. 317). Yet, although he is relatively content with his lot in life, he is not allegiant to the political regime. Nor is he alienated. He is rather, according to Lane, ''divorced.'' Dempsey simply lacks an explicit awareness of the political world to which he can either positively or negatively respond. Therefore, for him the political landscape is *terra incognito*: ''It is a timeless, changeless world for Dempsey, a world he never made, and beyond his control, and to a large extent beyond the control of even the 'higher-ups' '' (Lane 1962, p. 318).

Hodder. In Smith, Bruner, and White's in-depth study of the opinions of ten male Americans toward the Soviet Union during the late 1940s (1964, pp. 196–203), nine possessed IQs above 120 and, according to the authors, fell within the upper five percent of the population in cognitive ability. There was one exception: ''Hodder provided us with our best opportunity to examine opinions in a person little given to reading, reflection, or the more abstract operations of the mind. . . . At a concrete level he functioned effectively, showing good common sense and practical judgment. It was in the realm of abstraction that his limitations were most marked.'' Accordingly, we have placed Hodder in the narrowly abstract category; yet in view of his above-average IQ of 113, he must be placed in the upper reaches of that category.

Hodder was a factory operative and a union member: ''He was casual about his union participation, little interested in long-range plans for social betterment, quite contented with his position. . . .'' Although he had had a drinking problem, ''Hodder's genial, unambitious way of life allowed him to extend his philosophy to specific issues without finding many objects of bitterness''—the Russian Communists representing one of the few exceptions by posing as a threat to ''his generally satisfactory world.''

More allegiant than either alienated or divorced, his political views still lacked saliency or perspective. Indeed, the authors of the *American Voter* (Campbell et al. 1964, ch. 9) cite the case of Hodder as an illustration of the largely nonideological character of the political attitudes held by most American voters. Ironically, they also cite Hodder's above average IQ to demonstrate that intelligence is not a factor in producing this political phenomenon. An examination of the other cases in *Opinions and Personality*, however, suggests that individuals who are narrowly or functionally abstract and who are, from their own vantage point, content with their life situation, will lack both the cognitive skills and the situational impetus for political thought of an ideological character. A case in point immediately follows.

Lanlin. Smith, Bruner, and White (1964, ch. 7) view Lanlin as being especially close in his identification to the values and aspirations of the middle class. Compiling an uneventful, mediocre record through high school and business trade school, holding a number of jobs subsequent to graduation, including a frustrated period of nineteen years as a clerical worker, he had become by the time of the study a salesman of heating, cooking, and household appliances. "The process of selling," according to the authors, "gave him his greatest satisfaction." And the shift from clerical work to selling prompted these closing words in an autobiographical account prepared for the study: "Most important lesson I learned in life—get into work you enjoy."

Lanlin, according to his *Wechsler-Bellevue Adult Intelligence Test*, was "a man of superior intelligence" with an IQ of 120. Yet, "the subtests of the *Wechsler* as well as the *Vigotsty* and the *Rorschach* portrayed a man who used his endowment best in detailed, concrete, and ordered situations that did not require the handling of higher abstractions."

The foregoing represents the basis for placing Lanlin in a functionally abstract category (3, iii) in Figure 6.2. In any event, he does appear by and large satisfied: "His approach to the world was a smiling one" and his views toward public affairs in general were "strongly marked by gradualism and a rather unquestioning optimism." But his perspective tended to be limited to "the world of family, work, and the 'face-to-face' local community." Hence, opinions in more remote matters, such as "Russia and Communism were not intense," for "the main stream of his energies was directed toward matters much closer to his home and to his work." It was "only in so far as Russia and Communism impinged upon these more personal spheres that our subject felt intensively on these issues." Russia then became viewed "almost exclusively from the outside" as a threat to world peace. And that perception was drawn personalistically, for the authors note "Lanlin's incapacity to grasp or even consider possibilities of conflict based upon abstract principles."

Chatwell. An IQ of 142 placed Chatwell within "the highest one percent of the general population," as well as above any of the other subjects in the study (Smith, Bruner, and White 1964, ch. 6). Indeed, "only Chatwell seemed to operate with pleasurable ease in the realms of abstract thought." According to the *Vigotsky* test, he manifested the "highest level of abstract thinking." This capacity found suitable challenge in his pursuit of a legal career through attending law school and clerking in a small firm of patent attorneys.

In Chatwell, the striving for self-selection was both well developed and effectively realized: "A man who displays initiative and enterprise, who is gifted and experienced in rational mastery, and who rises to challenging situations with colorful self-dramatization, can take a good many things in stride. At twenty-eight,

Chatwell had achieved an effective adjustive technique, a 'style of life' which had proved successful under a variety of circumstances.'' According to the authors, ''His central pattern of successful strivings particularly colored his orientation and policy stand toward Russia. He advocated an active course of argument, persuasion, negotiation, and human management.''

Chatwell had a highly informed and differentiated view of Russian Communism: ''Chatwell's opinions were in this sense an expression of his mental abilities and of other relevant traits, quite apart from their content and their relation to the satisfaction of his strivings. The outstanding fact is his intellectual competence, which constantly encouraged both argument and reflection.''

The Trapped

If individuals feel stifled, they will want to escape, avoid, or strike at that which confines them. The politics of the trapped, therefore, lean toward the radical, even the revolutionary. This tendency is most pronounced in the highly abstract, where it is likely to take on an ideological form. The functionally and narrowly abstract make better candidates for the True Believer. The latter are also more apt to be influenced by changes in their immediate environment, so that a mass protest movement generated by adverse economic conditions may melt away in the sun of a returning prosperity, leaving an intellectual leadership as disaffected as ever, but perhaps more cynical and prone to retreat into privatism.

Daniel. Since the two studies used did not provide a narrowly abstract individual who was clearly trapped, we begin at the functionally abstract level and compare two individuals, one of whom is Hodder from the satisfied category, of differing abstract capacities in similar job situations. ''Although he was a factory operative and a union member,'' Smith, Bruner, and White (1964, pp. 189–96) observe, ''Sam Hodder contrasted in almost every other respect with Ernest Daniel.'' With an IQ of 128, for Daniel the ''job was by no means commensurate with his mental ability.'' In contrast to Hodder, union work ''soon became the center of his life'' and he rapidly became vice-president in the state organization. Nonetheless, ''he was not fully satisfied either by his own progress or by that of his union as a whole.'' Meeting defeat in a number of elections, he never succeeded in becoming a paid union official.

Daniel was one of the few in the study to hold a sympathetic view of Soviet communism: ''His opinions served to reinforce the image of himself as a member of the labor class and to strengthen his identification with the labor movement.'' As a result, he looked favorably on the welfare state aspects of the Soviet regime and ''expressed his preference for an arrangement under which an unskilled workingman would be given the chance to develop to the top of his capacity, and would in any event be protected from financial insecurity.''

Sullivan. Born of poor Catholic parents and educated only to the second year of high school, Sullivan nevertheless ''read omnivorously'' and registered an IQ of 128, including a verbal IQ of 134 (Smith, Bruner, and White 1964, ch. 8). It is plausible, given his self-education; his subsequent unemployment and alcoholism during and subsequent to the 1929 depression, and his marital problems, which ended in a divorce, that his ability remained understated. Unlike the other subjects (Chatwell excepted), "Sullivan's liking for large ideas was manifested in every topic he discussed." One of the best informed of the ten, he was without question the most ideological in his opinions, and his "communist beliefs were his more salient characteristics." A part-time freelance journalist, Sullivan was "a Marxist and a non-party Communist." A man deeply frustrated by life, the authors conclude, his attitude "helped him construct a world in which his life prospects did not appear catastrophic."

The Overwhelmed

If the trapped strike out at the status quo, the overwhelmed are wed to it. They have invested so much time, talent, and energy to get a position that demands every bit of their ability, and then some, that they can hardly repudiate either their efforts or the larger system of which those efforts seem to be a part. If the trapped turn radical, the overwhelmed lean toward the reactionary. Straining under the present system, yet wanting no radical change, they may yearn nostalgically for the illusory "good old days" when times were simple and honest effort was appreciated.

Rapuano. A significant finding in Lane's study (1962, p. 429) of the American "Common Man" is the distance of his subjects from Marxist thought: "The dialectic employs a time sweep outside his ken or interest; it is beyond his pragmatic inclinations." Lane opens his study with chapters illustrating that freedom has for many in his sample become a burden rather than a source of opportunity ("The Burden of Freedom") and equality a state to be feared out of feelings of inferiority rather than to be pursued out of ambition ("The Fear of Equality"). Thus Ferrerra, a wholesale shoe salesman, feels "constantly bombarded" with "tiny messages." Such individuals, Lane (1962, pp. 343–45) concludes, "have already escaped from freedom into obligations almost beyond their capacities."

Rapuano, in addition to Ferrerra, provides the clearest illustration of the overwhelmed at the narrowly abstract level. Asked if he does not feel sometimes that listening to all points of view on a subject is too confusing, Rapuano replies, "Oh, certainly, Oh, of course. Sure. . . . Maybe they don't want me to understand them—I don't know." Such responses lead Lane (1962, p. 110) to conclude that many in his sample "haven't the equipment and the background to understand."

Lane (1962, p. 181) is impressed with the way Rapuano "selected a career, rather than falling into an available vacancy, and the way he has built up his role in the packinghouse so that it now represents an important part of the operation." Yet his job is too much for him: "He must keep track of packinghouse supplies, endlessly worrying about innumerable items and paper forms. One might say of him that his malaise is exactly the reverse of the alienated man; he cares too much, he is too involved."

Rapuano's utopian dream appears confined to a South Pacific retreat; and if he could remake the world, he would simplify it. "What kind of government would there be in a perfect society?'' Lane (1962, p. 20) asks. Rapuano replies, ''Well, there certainly would be a simple government. . . . There wouldn't be too many people making the laws or governing the people, for one thing.'' And quite consistently, Rapuano thought Senator Joseph McCarthy ''was a great man.'' In accounting for these political views, Lane (1962, p. 110) writes: ''Since confusion about himself, combined with ambivalence about political matters, and uncertainty about his occupational future are constantly framing questions in his own mind, and have made him chronically worried and badly ulcerous, he cannot bear more confusion and indecisiveness in his world.''

Clark. According to Smith, Bruner, and White (1967, pp. 203–10), Clark's IQ of 125 placed him ''in the top 4 or 5 percent of the general population, but in the less structural *Rorschach* his stereotyped, constricted performance might have been given by a person of no more than average ability.''

Clark's singleminded, uncreative mentality is well reflected in his efforts at self-selection. An ''accountant who quite early had been forced into a hard psychological bargain with life and had lived ever since within the terms of the contract,'' Clark worked part time beginning at age thirteen, while continuing his education through two years of a business college. ''It was a stiff grind,'' but he had attained a ''secure white-collar position in a stable concern'' and ''a home for his wife and two children in a 'better-than-average' neighborhood.'' The two strongest preoccupations in Clark's current life, according to the authors, ''seemed to be his job and his son. In his job he had risen to the head of his department, with three men under him. He had reached the ceiling of probable advancement.'' On the one hand, ''he found satisfaction in precise work with figures,'' but on the other, ''he complained of having to work under pressure'' and of the necessity of directing his subordinates. ''Nevertheless, he did not intend to do any thing that would imperil such security as he enjoyed . . . no freely operating motives urged him toward greater power or greater responsibility.''

The combination of a narrow conceptual focus, on the one hand, and overtaxing occupational responsibilities, on the other, produced in Clark attitudes that were apolitical:

> Having small resources for outgoing interests and attachments, he ruled out large areas of public concern which did not directly touch his daily life. About world and national affairs he cared hardly at all. Things were bad but you couldn't do anything about it. Even community affairs drew little of his interest.

His opinions about the Soviet Union fit this picture: They ''were all unfavorable but of low intensity and little differentiation. . . . He perceived Communism as a great evil—but very far away from the accounting department of his company, and from his home in the better-than-average neighborhood.'' Communism was a ''great evil,'' it seems, because it threatened, however vaguely, the most important values and concerns Clark was to pursue.

WOMEN AND ALIENATION

> There is a huge price to pay for success. Any woman who tells you that she can hold down a high professional position (not a job but a career), raise children, be a good wife with a home and social life, and not suffer for it is not telling the truth.
>
> For me, the price has been well worth it. A hundred times so! It is almost for my make-up to feel I am making a contribution, and this carries over to my career, my family, and the various boards on which I serve.
>
> —Nancy W. Collins, *Women Leading*

> The trapped housewife was replaced by superwoman, who somehow managed to rear her children while catapulting ahead in a career. Now, with legions of superwomen worn out and discontented . . .
>
> —*The New York Times*, September 23, 1988

This section is more sketchy because, for women, there is no intensive case study such as Smith, Bruner, and White's *Opinions and Personality* (which included only men).

Entrapment

Clearly, women in general fit the category of entrapment from the feminist perspective. A feminist collection entitled *The Trapped Woman* (Figueira-McDonough and Sarri 1987) illustrates this position, for as the editors write in their first chapter, women are to be thought of as a "minority group" because "of the individual and collective discrimination that they experience and because of their inability to resist being either thought of or treated as inferior" (p. 11).

Our analysis suggests, however, that some women (as with men) are more likely than others to feel entrapped. In particular, the more cognitively capable the woman, the more likely that she would not find an environment congenial to her ability. This likelihood has been especially strong historically among the intellectually abstract, for they will almost never be fully satisfied with any environment precisely because of their searching, critical intelligence.

Some female intellectuals and artists implicitly recognize the force of these statements. Walker (1977, p. 95), for example, calls attention to the example of Phillis Wheatley, a slave, who "had she been white, would have been considered the intellectual superior of all the women and most of the men in the society of her day." This (ironically somewhat sexist) assertion is then placed within the context of Woolf's treatment of the talented woman in *A Room of One's Own*. Walker (1977, p. 95) cites Woolf, at the same time interjecting her own comments to make Woolf's statements directly relevant to Phillis Wheatley:

> [A]ny woman born with a great gift in the sixteenth century [insert *eighteenth century*, insert *Black woman*, insert *born or made a slave*] would certainly have gone crazed, shot

herself, or ended her days in some lonely cottage outside the village, half witch, half wizard [insert *Saint*], feared and mocked at. For it needs little skill and psychology to be sure that a highly gifted girl who had tried to use her gift for poetry would have been so thwarted and hindered by contrary instincts [add *chains, guns, the lash, the ownership of one's body by someone else, submission to an alien religion*], that she must have lost her health and sanity to a certainty. (emphasis in original)

De Beauvoir, in the classic *The Second Sex* (1953), generalized about women as being metaphorically in a "prison" (p. 717), but de Beauvoir is especially attuned to the acute sensitivities of the "independent woman—and above all the intellectual" (p. 685). Interestingly, de Beauvoir describes her own emergence as a feminist not as a result of abstract theory, but directly as a problem in self-fulfillment, that is, in self-selection:

When I first met Sartre, I felt I had everything, that in his company I could not fail to fulfill myself completely. Now I reflected that to adapt one's outlook to another person's salvation is the surest and quickest way of losing him.

But why, when all is said and done, should I have felt such qualms of remorse? I was certainly not a militant feminist: I had no theories concerning the rights and duties of women. Just as previously I had refused to be labelled "a child," so now I did not think of myself as "a woman": I was *me* . . . to accept a secondary status in life, that of a merely ancillary being, would have been to degrade my own humanity; and my entire past rose up in protest against such a step. Obviously, the only reason for the problem presenting itself to me in these terms was because I happened to be a woman. But it was as an individual that I attempted to resolve it. (1962, p. 71; emphasis in original)

Later, de Beauvoir developed a more abstract feminism and, for example, advocated a "general refusal" to participate in the "star system," by having feminists not sign their names to their writings, thereby refusing "the idea of competition, masculine glory, ambition, and fame." But when Betty Friedan challenged her as to whether she was "no longer going to write books under your own byline," de Beauvoir, responded, "No, of course not, because I was formed differently" (Friedan 1985, p. 310).

The original impetus for the writing of *The Feminine Mystique* by Friedan came as a reaction to her analysis of a questionnaire of her college classmates, fifteen years after their graduation from Smith. She observed the "schizophrenic split" these women faced between the abilities that their sophisticated educational background presupposed and their expected conformity to a passive intellectually condescending feminine role (Friedan 1974, p. 11).

Indeed, Freeman, in *The Politics of Women's Liberation* (1975, pp. 35–37), notes that college women have in general "greater expectations and lesser opportunity for realizing them" and that undoubtedly the college-educated woman compose a disproportionate number of the women's liberation movement, as suggested by the fact that over four fifths of *Ms* subscribers have some college education, compared to roughly one fifth of the total female population.

The literature on the women's liberation movement in the United States is replete with specific examples of entrapment. Evans (1979, p. 234), for example, reprinted the Student Non-Violent Coordinating Committee (SNCC) position paper of November 1964, on "women in the movement," which states "many women in the movement are not happy and contented with their status" and "much talent and experience are being wasted by this movement when women are not given jobs commensurate with their abilities." Altbach (1971, p. 143) describes the case of Naomi Weinstein, who "after completing her postdoctoral fellowship in mathematical biology at the University of Chicago (after obtaining her Ph.D. in psychology from Harvard in two and a half years)" was rejected for an academic position at 12 institutions because she "lacked the qualifications": "It was enough to make one join Women's Liberation, which she did, in 1966."

More generally, Klein (1984, p. 115, 119) observes that women who have felt discrimination are more prone as a result to feminism and that "women in a traditional role, such as being a housewife, develop a feminist consciousness if they believe that traditional arrangements hinder their personal opportunities to achieve goals and gain status." In individual cases of women such as Virginia Woolf, Alice Walker, Simon de Beauvoir and Betty Friedan, their highly abstract mental capacity is easily established and therefore their presumed increased sensitivity to discrimination. But what is typically missing in more general or statistical observations is a knowledge of the individual level of the capacities of the women discussed. In light of our analysis, it would be the more intellectually gifted housewife who would feel trapped (cf. Friedan 1974), and if such a point seems obvious, the absence of any relevant data either to support or disprove it seems glaring.

The Satisfied

Despite the disarray of literature on women's vocational satisfaction (Betz and Fitzgerald 1987, p. 186), we will concentrate on a number of individual case studies of women who appear to be pragmatically abstract and reasonably successful. One such woman, Nancy Graves, grew up in a household where it never "occurred to anyone on either side of the family that a woman could be or ought to be 'a mere housewife' " (Ginzberg and Yohalem (1966, p. 13). Born in 1918 in a small midwestern town where her father was a high school principal and her mother an organist and both were college graduates, she completed her own undergraduate education at the state university. After obtaining a master's degree at Columbia, she eventually entered college and university administration, at one point becoming a dean of students at a small liberal arts college. At the time of Ginzberg and Yohalem's study, she was in student personnel administration at a new branch of a state university.

Graves apparently never married, but her life, especially in professional terms, had gone well: "I would have been a very unhappy woman without a higher education and a career because I have scholarly ability and have wanted to develop it as far as I could.'' She states that she is ''presently well satisfied with my position'' and that ''I believe I am working at a level that is commensurate with my abilities.'' Thus, she seems to fit the ''satisfied'' category at the pragmatically abstract level.

Graves's position on feminism seems noncommittal: ''I have not been particularly concerned either with assistance from others in the profession or with obstacles as a consequence of my being a woman. Perhaps my attitude and expectations were naive, but I felt that I had to earn assistance and overcome obstacles by my own performance'' (Ginzberg and Yohalem 1966, pp. 17–19).

Gretchen Minnhaar was born in Argentina, where her father was a medical internist and her mother was active in ''humanitarian endeavors'' (Minnhaar 1974, p. 26). After her marriage to a medical doctor, she accompanied him to the United States (Detroit) where he sought further training and where she continued her college education, which she completed in Argentina following their return a number of years later. As a highly trained specialist, however, her husband found it desirable to pursue his career in Detroit, where she secured a job in one of the largest architectural firms in the country.

Minnhaar eventually became a professional architect in Grand Rapids. She has found satisfaction in her marriage and family (one daughter) and in her professional career. She has few regrets: ''To work outside my home in a liberal profession implies extra effort but never a sacrifice, since this is what I want to do.''

Minnhaar, like Graves, is not an ardent feminist: ''Now, unwilling, I seem to be part of the Women's Liberation Movement, a fact that I resent because I refuse to be bunched under the heading 'women' and then see all the things that I must be or do'' (Minnhaar 1974, p. 31). (Note the ironic parallel here with de Beauvoir with respect to the resistance to being categorized as a woman, but this strong individualism becomes replaced in theoretical formulation, if not practical action, in de Beauvoir.)

Both Minnhaar and Graves seem to say that they made it on their own with great effort and without major barriers, and other women can do so as well on their own. This outlook is perhaps not unlike that of successful Americans in the past who have risen individually from modest backgrounds and, as a result, resisted radical thought—again excepting in many cases the intellectually abstract.

In the volume from which the Minnhaar experience is taken, *Women and Success* (1974), there are eleven other professional cases, all of them successful and, therefore, vocationally satisfied. All, like Minnhaar, appear to be pragmatically abstract (based on the inference from the choice of professional careers: pediatrician, chemist, meteorologist). Their commitment to feminism is, on the whole, not pronounced. This assessment is shared by a feminist sociologist, commenting on the cases in the same volume (Hochschild 1974, p. 199): ''The women whose life stories appear here are at best ambivalent about the women's movement.'' Indeed, Friedan (1985, p. 19) discovered that *The Feminine Mystique* was ''strangely threatening'' to some career women who had already made it ''as freaks or exceptions'' in the male world.

According to Hochschild (1974, pp. 194, 196), who also refers to "the exceptional woman," "Those at the very top of society and at the very bottom are probably the least likely to recognize women as a minority group": "At the bottom, it is because the woman's husband is hardly better off, and at the top it is because the woman often assimilates to the male elite at the price of identifying with males and looking at women, in part, as a semi-alien group," which tends to happen "to virtually every ascending and assimilating group."

These general conclusions have a rough validity, but they must be qualified by our individually based analysis. Those at the bottom who are intellectually gifted, such as Alice Walker, are likely to be especially sensitive to sexist and other discrimination. Even if such exceptionally gifted rise to the top, their intellectual sensitivities are likely to retain a radical edge. The "excepional woman" who is more pragmatically than intellectually abstract may find greater satisfaction and hence may moderate her views accordingly.

The Overwhelmed

The rise of a postindustrial, high tech society in the West has increased the numbers of overwhelmed in the work force. Presumably women are not exempt from this trend, as opportunity, if not actual demand, for their services in sophisticated occupational lines increase. In the business world, Loden (1985, p. 212) writes, "There is little doubt that rapid change and growing complexity have placed modern executives under increasing amounts of stress." Citing scientific evidence that women have "more built-in hormonal protection to withstand chronic stress with less reactive physiological consequences," she suggests that women may be less susceptible to succumbing to the stresses of a job situation. This, of course, would not apply to women who suffer from premenstrual syndrome (PMS) (cf. Walsh 1987, Part II). In the absence of more definitive studies in this area, no conclusions can be drawn. It must be readily acknowledged, though, that significant biologically based differences by sex may exist.

No doubt women feel stress in their work. In a sample of working women who are high achievers—160 women listed in *Who's Who of American Women*, Collins, Gilbert, and Nycum (1988, p. 9) conclude: "Almost all of the respondents to our survey report some stress in their lives, but less than 15 percent reported feeling in a state of almost perpetual stress." They continue: "Many professional women we know leave their offices with stuffed briefcases, fight the traffic to get home, unwind with a drink and a late meal, and then work for a few more hours before collapsing into bed. As one woman wrote, 'Ever feel stressed? Of course, it comes with *all* high level jobs!' " (emphasis in original).

It is logical to assume that stress is affecting more women as more opportunities for high level advancement have become available. The number of women reported to feel in a state of almost perpetual stress in the Collins, Gilbert, and Nycum study should be highlighted, not downplayed. Almost 15 percent of top women achievers feel perpetually stressed. That is simply not a negligible proportion. Future research needs to examine how this figure compares to women in a previous generation and to men in this one.

Undeniably, sexual discrimination in employment still remains—the "glass ceiling," above which women may find they cannot rise in an organization. Bardwick describes this phenomena in *The Plateauing Trap* (1986), although she does not restrict it to women alone. Accordingly, she (1986, p. 30) denies that plateauing can be related to the Peter Principle: "*That is totally wrong*. Today plateauing and *The Peter Principle* are the reverse of each other" (emphasis in the original). Yet one wonders if she does not protest too much in suggesting that everyone in the work force who has reached the limit of their occupational climb is trapped rather than satisfied or, at least in some cases, overwhelmed. She acknowledges that in her research most people made precisely the latter connection; indeed, she notes that many plateaued people find it hard to avoid thinking of themselves as being part of the Peter Principle (Bardwick, p. 32): "Well, perhaps some of them are not wrong. Of course, there's always some truth to the Peter Principle" (p. 33).

As an example, Bazelon Goldstone (1988, p. 203) became the head trader for currency options at a large Wall Street concern at twenty-seven: "The strain of expansion, with the resulting overflow of operational and management problems, combined with insufficient sleep, were beginning to take their toll. Gone was the confident, optimistic, bright-eyed trader of the previous year. In her place was a frothing maniac: unsure, overworked, exhausted." In short, overwhelmed.

Even those who are satisfied in their occupational role may find additional, more traditional household chores to be onerous. Thus a sculptor and art professor at Wellesley College, Alice Atkinson Lyndon (1977, pp. 36–37) writes:

> I want to emphasize that I am not unwilling to perform tasks that are generally considered feminine, but I am unwilling to sustain what becomes for me in the instance of cooking a particular kind of work overload. I work with my own, or with students' formal juxtapositions all day long; especially since I began teaching sculpture as well as doing it, I find that when I get home from work I am depleted of the creative energy it takes to "arrange" foods to make a meal. For me, food combinations such as walnuts on broccoli—like additive sculpture—involve presentations of objects in terms of their color, shape, and spatial arrangement. . . . If I feel many times that I cannot make such a presentation, this is not a matter of feminist protest but the result of "shape-juxtaposition fatigue."

A point of interest here is that the objection to traditional household work is not feminist or ideologically based, but practically based on the grounds of work overload. This comment returns us to the possible biologically based differences between the sexes that Rossi (1977, 1987), among others, refer to with respect to child bearing and caring. Specifically, the superwoman syndrome, or the difficulty that a woman has in "having it all"—marriage, family, and career—clearly does not exist for men.

The superwoman syndrome can be interpreted differently depending on viewpoint. A feminist perspective, for example, tends to see the "superwoman" as trapped by culturally caused sexism. Thus, Fox in *The Trapped Woman* (1987, pp. 206–07) asserts that along with "deprivations in occupational location and rewards, traditional definitions of male, female and sex-appropriate domains create other tensions, strains, and institutionalized deviance (no-win, double-binding) criteria for women at work." Fox notes that "despite the dramatic increase in women's labor force participation, the location of household responsibilities has, in comparison, changed very little" so that "women's family roles impose onerous demands and constraints—both at home and at work." These demands are indeed real, but whether the traditional division of labor that imposes them is entirely cultural is more problematic.

A more optimistic feminist position is also possible. Loden (1985, pp. 208–10) sees the role of superwoman as being, at least potentially, self-fulfilling. Acknowledging that this role has "in many respects" been "unfairly imposed on working women in today's society," Loden nonetheless argues that it offers "many satisfactions and rewards." Loden cites the conclusion of Dr. Lois Verbrugge of the University of Michigan: " 'The more roles women have, the better their health. Women with three major roles (employment, marriage, parenthood) tend to have the best health status.' " This position rests on two assumptions, explicit and implicit, but both probably unrealistic. The explicit assumption is that "adequate family, employment, and community support" is available, for example, a sexually egalitarian domestic division of labor and community-sponsored day care centers. The implicit assumption is a uniformly high potential capacity on the part of all women.

Another perspective is to view the would-be "superwoman" as overwhelmed. Comparing her male and female patients, the therapist Georgia Witkin-Lanoil (1984, p. 5) concludes: "The women spoke of many more stresses that were long-term and largely beyond their control—the two factors that make stress dangerous to psychological and physical health! They spoke of unequal pay and unequal say. They spoke of double duties: housework and work-work." Or as Eichenbaum and Orbach (1988, p. 30) put it: "Women have learned that achieving status in the work place and at the same time remaining committed wives and loving mothers is often unbearably demanding. . . . The enormous stress of these new expectations is becoming obvious as the incidence of heart attacks and lung cancer go up for women."

What these differing conclusions point to is the need for more empirical study. In particular, in-depth studies in which the individual capacities of each person is explicitly treated are required.

BIOLOGY, HISTORY, AND POLITICS

We may recall that individuals may change their social positions over a life time. If such individual changes are collectively numerous, similar in direction, and more or less simultaneous in time, their larger societal impact is ensured. They then become a critical part of the historical process.

In order to explore the political ramifications of such large scale societal change, Galbraith's analysis (1967) of the changing nature of the labor force throughout the modern era will be employed. Of course, we must keep in mind that the life work of an individual is by no means the most important component of his or her environment, but it here it serves, as a convenient environmental measure.

Galbraith (1967, p. 248) describes the work force at the start of the Industrial Revolution as follows:

> In the early stage of industrialization, the educational requirement for industrial manpower was in the shape of a very squat pyramid. A few men of varying qualifications—manager, engineers, bookkeepers, time keepers and clerks—were needed in the office. The wide base reflected the large requirement for repetitive labor power for which even literacy was something of a luxury. To this pyramid the educational system conformed.

Now, the situation has changed radically, at least for urban societies in the West and in Japan:

> By contrast the manpower requirements of the new industrial system are in the shape of a tall urn. It widens out below the top to reflect the need of the technostructure for administrative, coordinating and planning talent, for scientists and engineers, for sales executives, salesmen, those who worked in the other arts of persuasion and for those who program and command the computer. It widens further to reflect the need for white-collar talent. And it curves in sharply toward the bar to reflect the more limited demand for those who are qualified only for muscular and repetitive tasks and readily replaced by machine.

Moreover, "the top of the urn continues to expand while the bottom remains the same or contracts" and "to this change the educational system responds" (Galbraith 1967, p. 48). Nevertheless, this process should not be viewed as inevitable or unidirectional. By 1970, the American economy experienced a substantial cut back in its demand for "techno-structure" specialists and in its support for graduate education.

The images presented in the previous excerpts can be contrasting historical metaphors for the labor force. In visualizing the "tall urn," we obtain the relationship between the distribution of intelligence and its occupational demand as described by Cattell (1965, p. 40) in Figure 6.1.

Cattell is concerned with the economic ramifications of the supply and demand of intelligence in society, and his figure clearly shows the greater demand for (trained) intelligence than its supply and the subsequent relationship to earnings; that is, the brighter get better (more skilled) jobs and make more money.

Our focus, however, is on the psychological and political ramifications of an "under supply" of intelligence (or abstract ability). If Cattell's suggested distribution of intelligence and its "undersupply" roughly mirrored American society for

Figure 6.1
Psychology and Economics: Relation of Mental Capacity Supply and Demand to Income

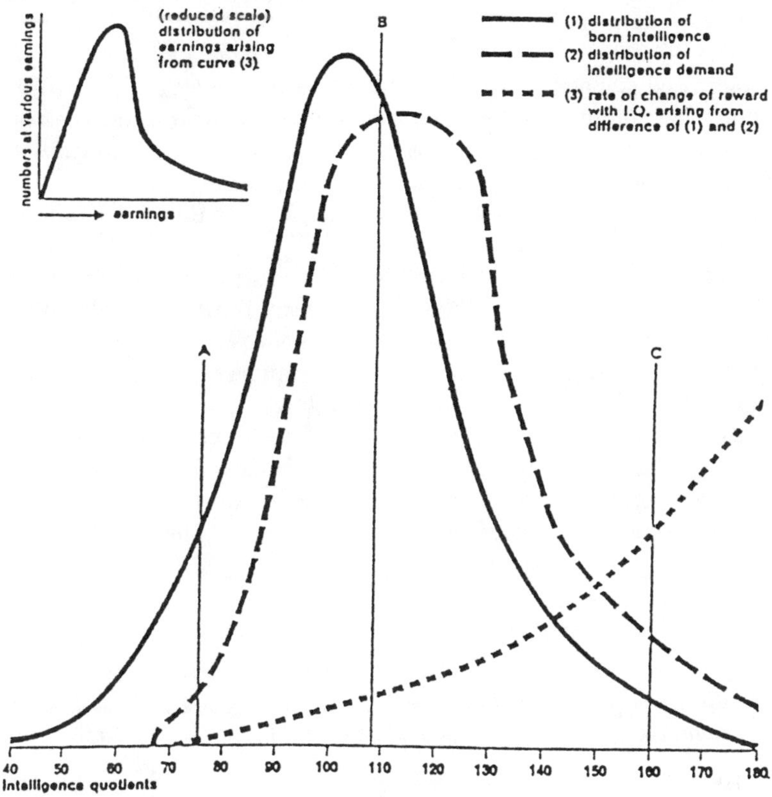

A at IQ 75 supply greatly in excess of demand, therefore negligible earning power; B at IQ 108 supply equal demand, therefore normal earning power; C at IQ 160 demand greatly in excess of supply, therefore disproportionate earning.

Source: R. B. Cattell, *The Scientific Analysis of Personality*, Chicago: Aldine, 1965, p. 40.

the period of the mid-1960s (during which he wrote), then the numbers of the overwhelmed should be increased. This is supported by Keniston (1968, pp. 369–70), writing during the mid-1960s:

Until recently, the supply has been adequate to the demand: enough young men and women in each generation possessed enough ego skills to meet the growing demands of society. Indeed, America may have earlier possessed an untapped reservoir of "ego talent," of men and women who were functioning at less than peak performances, whose potentials were not used by their current lives. But since the Second World War, there are increasing signs of a growing shortage of people with the "right" kind of personality. Specifically, we find an ever more urgent and deliberate effort to "produce" men and women with technological egos and to prevent the "human failures" who lack the characteristic virtues of our society. A glance at the employment pages of the *New York Times* or the *Scientific American* will confirm the scarcity of those with the most highly developed technological skills. (emphasis in original)

This increased demand for skills, concludes Keniston (1968, p. 370), means that "more and more men and women are pushed to perform at the peak of their ability" and that a distinction must be made "between those who *cannot* meet the demands of their society, and those who *choose not* to do so." Keniston thus anticipated the soon-to-be-published Peter Principle, even if he failed to recognize its underlying biosocial basis.

Returning to Galbraith, we will use his "very squat pyramid" as a metaphor for the preindustrial work force. Although intelligence is still genetically influenced and normally distributed for a society, this time a restricted demand exists for its occupational use. Figure 6.2 describes the relationship between the demand for and the distribution of intelligence that follows.

Here entrapment characterizes many in the labor force. But as the nature of the labor force changes through time and the demand (and reward) for intelligence increases, the incidence of entrapment should decrease. At some point, more in the work force may become overwhelmed than trapped. The suggested changes

Figure 6.2
Relationship between the Distribution of Existing Intelligence and the Distribution of the Demand for Intelligence at the Outset of the Industrial Revolution

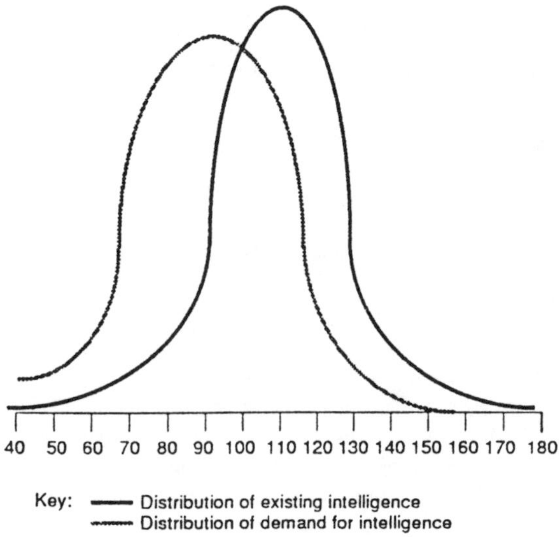

in the numbers of trapped and overwhelmed as societies move from pre- to postindustrial status are shown in Figure 6.3.

These figures are presented primarily for heuristic purposes, for their inability to distinguish more carefully among those who are either trapped or overwhelmed results in the loss of critical information. For all its flaws, however, this analysis may point the way to more precise study in the future. For the present, it may suggest not only the existence of critical political differences between pre- and postindustrial societies but also their underlying basis.

Preindustrial Society

In considering contemporary industrial nations theoretically at the outset of their economic development, the most obvious point suggested by an analysis of Figure 6.3 is the potential in the population for revolutionary change. If numerous individuals in a society feel trapped, they form an environment that accentuates their frustration. This situation is further aggravated if potential elites—for example, many of the highly abstract—are denied access to positions of power and influence.

In part, the failure of a society to provide suitable opportunities for its members may often be ascribed to the narrowly possessive and insecure character of its privileged classes. Such was true in France during the Ancien Regime. A founding noble may indeed possess aristocratic virtue, but as a result of genetic diversity

Figure 6.3
Changes in Proportion of Trapped and Overwhelmed as Societies Move from Pre- to Post-Industrial Status

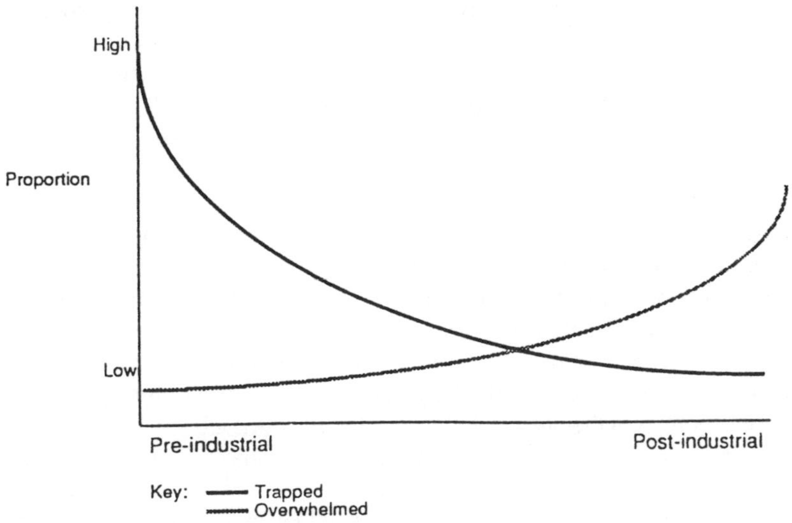

and perhaps excessive in-breeding (cf. Darlington 1971), some of his successors are apt to confirm the Peter Principle upon birth. Such a circumstance occurred with the advent of Louis XVI. That unfortunate monarch, according to Lefebvre (1959, p. 23), "though jealous of his authority, was lacking in will; honest and well-intentioned, he was far from being a great mind." Lefebvre goes on to argue that "among the immediate causes of the Revolution the character of the king and queen must be included. It is scarcely doubtful that events would have taken a different turn if the throne had been occupied by a Henry IV or even Louis XIV."

Individuals within the privileged classes who are overwhelmed exacerbate a revolutionary conflict by stubborn, uncompromising stands. In a sense, the great revolutions in history have been struggles between the trapped of the rising classes and the overwhelmed of the entrenched classes. On the basis of his study of revolution, Sorokin (1959, p. 372) concludes:

> The prerevolutionary society in ancient Rome and Greece, in the Middle Ages, before the English Revolution of the seventeenth century, before the Russian Revolution, before the great French Revolution, exhibit conspicuously the abnormal defectiveness of social distribution of its members. In the upper strata is found a great proportion of effeminate, idle, and incapable individuals who, according to their qualities should be put within the lower stratum. In the lower layers is an extraordinary accumulation of the energetic "climbers" barred from climbing to the upper position. The degenerated aristocracy is incapable of fulfilling its business of social control; the pushers underneath undermine the existing regime from the bottom. The result is a revolutionary explosion and its consequences.

The Postindustrial Society

If we assume genetic variability and a continuing normal distribution of such key genetically influenced traits as intelligence, then a postindustrial society, like its preindustrial predecessor, will be anything but monolithic. Everyone in such a society will not become uniformly educated, sophisticated, and competent.

To the contrary, the rise of scientific and technological advances by a small selected minority produces the "cultural lag" that Ogburn (1922) correctly observed over a generation ago and the "future shock" that Toffler (1970) has overgeneralized about more recently.

Thus, Drucker (1974, p. 32) observes the replacement of the manual worker by "the knowledge worker who puts to work what he has learned in systematic education, that is, concepts, ideas, and theories, rather than the man who put to work manual skill or muscle." Reich (1991, p. 3) discusses the rise of the "symbolic analyst," that is, those who "solve, identify, and broker problems by manipulating symbols." Take the current "robotics revolution" (Scott 1984) as a case in point. On the one hand, the need for unskilled labor is likely to diminish. On the other, "robotics is a highly multidisciplinary subject drawing on computing, mechanics, electrical engineering, mathematics, fluid engineering, metallurgy, control engineering, psychology, industrial sociology and so on." Thus, "however much as an expert one may be in any specific branch of technology, suddenly one finds one is a mere beginner in several of the other branches which are equally important to robotics!'' (Scott, 1984, pp. 27, 41).[2]

Economic progress increases the numbers of the overwhelmed throughout the entire populace. Neff (1985, p. 252), in his discussion of ''work stress,'' notes that the ''entire field of study and research is comparatively new, dating back less than 30 years.'' The white-collar ''Clarks'' and clerical ''Rapuanos'' are joined by many others.

Economic prosperity also permits the satisfied status of the ''Hodders,'' ''Lanlins,'' and ''Chatwells.'' Indeed, it is the successful self-selection of many members of society, especially those highly and functionally abstract, that create, à la Adam Smith, individual and thereby collective well-being. In this case, the numbers of the trapped decline to marginal proportions. Significant numbers of people will continue to be trapped, however, especially by the remaining snares of discrimination that especially affect the most gifted and thereby ultimately produce, in American society at least, minority and gender recognition movements. And the vagaries of circumstance, as in the cases of ''Daniel'' and ''Sullivan,'' will no doubt affect many others.

Nonetheless, the great majority of the ''masses'' will no longer be trapped. In fact, the most significant source of entrapment will be a perennially elitist one: the intellectually abstract, especially those among the young without responsibility felt to be commensurate to their talents. There will never be an ''end to ideology'' so long as this biologically inevitable grouping exists. The mundane standards of the many are bound to appear ''one-dimensional'' from the perspective of an idealist who, in spurning conventional social practices, may ultimately come to spurn the practitioners themselves. (In the Soviet Union, the intellectual dissident persisted across the generations, even though numbers remained relatively limited.)

Precisely because of self-selection, together with a natural increase in their numbers through the growth of the population, disaffected intellectuals who in the past may have been isolated and politically powerless find themselves clustered together on elite campuses or scientific and professional technostructures where their mutuality of interest and outlook becomes reinforcing and politically vitalizing.

The stage is thus set for a political polarization in which the preindustrial potential for a Marxist-style revolution becomes turned on its head. Instead of an overwhelmed elite confronting a largely entrapped mass, an increasingly overwhelmed ''silent majority'' rebuffs the utopian message of a radicalized intelligentsia. In this context, the generation gap is not between the young and the old. It is between the few young who are trapped and the old who are overwhelmed. Indeed, there is something of a generation gap in reverse between those young who are overwhelmed and those old who are intellectually abstract and subjectively entrapped. As long as the basis for entrapment and overwhelmingness remains, political conflict and the generation gap in some form will persist. But then so will life, history, and politics.

NOTES

1. While this proposition should not be pushed too strenuously, listen to the following testimony by Malcolm X and by Milovan Djilas, respectively: ''I don't think anybody ever got more out of going to prison that I did. In fact, prison enabled me to study far more intensively than I would have if my life had gone differently and I had attended some college'' (Malcolm X 1966, p. 184); ''Prison is a unique place in which a strong healthy man, able to resist, can discover his own capabilities,'' *New York Times*, May 16, 1972.

2. Emerging technologies have had their own impact on academia. Thus, Sternberg (1985, p. 110) speaks of the ''need to bring developments in cognitive research into the area as of current technology.'' As Bruner (1984, p. 63) states:

> It seems plain to me now that the ''cognitive revolution'' . . . was a response to the technological demands of the Post-Industrial Revolution. You cannot properly conceive of managing a complex world of information without a workable concept of mind. And it is probably not an accident that at the very time the cognitive revolution was occurring, America's Gross National Product was for the first time coming to be made up of the earnings of industries involved in communications and in the high technology of control, computation and management.

Chapter 7

Self-Selection and Sexual Politics

> The feminist movement has wanted to, as it were, really turn women into men. . . . [What] I find so disturbing . . . is the idea that family duties—rearing children, managing a home, loving and caring for a husband—are somehow degrading to women. And that's an idea which is very current in our society—that women are not going to find fulfillment until they get out there and start competing for a livelihood with men.
>
> —Pro-life activist, cited in Luker, *Abortion and the Politics of Motherhood*

> But there is a type of woman who cannot remain at home. . . . In spite of the place which her children and her family take up in her life, her nature and being demand something more; she cannot divorce herself from a larger social life. She cannot let her children narrow her horizon. And for such a woman, there is no rest.
>
> —Golda Meir, *A Land of Our Own*

> Perhaps the best approach is still the response of the eighteenth-century British writer Samuel Johnson, who when asked who was smarter, man or woman, replied, "Which man—which woman?"
>
> —Juanita Williams, *Psychology of Women*

This chapter applies the local-cosmopolitan dimension to the conflict over the Equal Rights Amendment. Especially relevant to this chapter are Schubert's masterly survey and critique of the literature on "sexual and feminist politics" (forthcoming) and Master's pioneering work (1989) linking in part—empirically as well as theoretically—neurobiologically based sex differences with corresponding differences in political cognition. Also noteworthy within the emerging area of biopolitics are works by Baer (1988), Caton (1985), Watts (1983), and Etkin (1979, 1985).

Before we tackle the ERA conflict, the local-cosmopolitan concept needs to be redefined given the likely biological basis for psychological differences between the sexes. In addition, this chapter proposes that (1) a relative tendency for women to become enmeshed in immediate social and family relationships will become reflected in their local-cosmopolitan orientation, especially the less cosmopolitan the woman; (2) a stronger maternal, in contrast to paternal, predisposition will influence female political thought and action, and finally (3) these differences became critically reflected by way of a dynamic self-selection process in the battle over ERA.

In examining these propositions, the wide range of individual variation that characterizes any population is stressed. Such a view is necessary when discussing group differences, as by sex, since frequently any inter-sex average differences are much smaller in range than intra-sex differences. The analysis concentrates on the crucial role of women activists in both supporting and ultimately defeating ERA.

Integrated into this final chapter are the differing mental capacities that characterized the participants. In particular, the most vocal advocates of ERA were closest of all to being intellectually abstract, with two important consequences. First, they felt a sense of entrapment from which stemmed their reforming impulse. Second, they were able to give intellectual expression to such feelings much more effectively than their opponents. In academic circles, for example, they totally dominated the debate. In the end, however, the feelings of entrapment of an elite few was not characteristic of the population as a whole.

SEX DIFFERENCES AND THE LOCAL-COSMOPOLITAN DIMENSION

Chapter 4 elaborates a continuum in local-cosmopolitan orientations for which the underlying basis is variation in the neurobiologically based capacity to envision space-time horizons. The latter involves an individual's capacity to extend his or her thinking over space and time and will likely vary by maturational age and by individual.

As presented in chapter 6, the more cosmopolitan space-time horizons involve metacognitive processes with an emphasis on global planning components, that is, a more general capacity to survey a situation, identify problems, and propose broad strategies to deal with them, in contrast to a narrow focus on specific tasks and more limited tactics. The relation of this metacognitive capacity or "metaintelligence" to self-selection and more specialized abilities is nicely suggested by Sternberg (1985, p. 50):

> Intelligence involves selection of environments as well as adaptation to them. To differing but usually nontrivial extremes, people are able to select the environments in which they

> live. . . . I believe that a key aspect of intelligence is the active selection, where possible, of an environment that enables one . . . to capitalize on one's strengths and to compensate for one's weaknesses. One can see these skills at work particularly among highly successful people in a given field of endeavor. Such people tend to share not any one particular ability . . . but rather a higher-order ability to capitalize upon whatever abilities they have in their work and to minimize the negative consequences of their weaknesses. Where possible, more intelligent people actively select environments as they perform more intelligently. To the extent that there is any aspect of intelligence that transcends particular environments, this aspect would appear to be critical.

Here we are concerned with the "higher-order ability" or metaintelligence that permits at least some individuals to transcend particular environments.

The local-cosmopolitan formulation outlined in chapter 5 presumably applies equally by sex, and represents a qualification of earlier formulations of the local-cosmopolitan dimension (Merton 1957; Dye 1966) that were implicitly environmentalist in basis. Let us examine the Dye formulation in particular, because it is in part the basis for a rejection by Arrington and Kyle (1978) of the thesis that a local-cosmopolitan division characterizes the fight over ERA. In their sample of activists in North Carolina, Arrington and Kyle (1978, p. 677) report that "the pro-ERA people were only somewhat more cosmopolitan" while "the anti-ERAs were slightly more locally oriented." (Note the small numbers involved in their individual categories—for example, thirteen men and fourteen women against ERA. In addition, the conclusion of Matthews and DeHart [1990, pp. 124, 153] show that for the women activists in North Carolina the ERA proponents—even the many with small-town or rural backgrounds—were "nonetheless cosmopolitan in outlook," whereas the opponents were "more likely to be oriented toward local than cosmopolitan values and experiences"). The basis for their conclusion was the application of the scale devised by Dye (1966, p. 241) to operationalize the local-cosmopolitan dimension. Although its five items in general tap the dimension, several possible difficulties should be noted.

The first item reads: "The most rewarding organizations a person can belong to are local clubs and associations rather than large nationwide organizations." Although the implication is clear that this dichotomy represents differing local/cosmopolitan preferences, it nonetheless overlooks the possibility that at least in some instances cosmopolitans may also be locals. In Adler and Bobrow's study of Aurelia (1956), a "well-to-do suburb of a mid-western city," for example, top level influentials are active and have influence "in both the local community and in the nation" (p. 97).

The second item contrasts national news with local happenings in their degree of interest, with the implication that a preference for the national over the local represents a cosmopolitan choice. Lane (1962, p. 137) concludes, for example, that "we are a nation of 'cosmopolitans,' not 'locals' " because most Americans are nationally, not locally news oriented. Yet this conclusion comes from his own in-depth study of working class men in New Haven, Connecticut, who are by no reasonable definition cosmopolitan. Indeed, a few are totally "divorced" from the larger political world and most are relatively apolitical. Most Americans may be "neighborhood" as opposed to "community" locals; in other words, the former are just as removed from local communitywide events as from the national scene. Furthermore, the national media may often personalize the news, precisely for audience appeal, with more abstract issues simply not being widely reported nor attended to by much of the public.

Finally, the fifth item reads: "I have greater respect for a man who is well established in his local community than a man who is widely known in his field but who has no local roots." Aside from its arguably sexist formulation, the item calls attention inadvertently to the fact that the scale is reflective of a male-oriented perspective, or at least one that fails to be cognizant of differences that might be based on sex.

It is not surprising that such a scale, devised in 1966, did not reflect possible differences by sex. The relevant, if still controversial theories of sociobiology, the pertinent research within neurobiology, and the influential developmental work of Carol Gilligan would not arrive on the scene for another decade. During the same time period there was a burgeoning of feminist literature.

Gilligan (1982), in her critique of what she sees as the male-oriented developmental theories of Piaget and Kohlberg, distinguishes between a male tendency to formulate moral questions in abstract terms and a characteristic female response that is rooted in a woman's apparently deeper enmeshment in close personal relationships.

Although Gilligan does not attempt to provide an underlying rationale for this dichotomy, it has been provided by sociobiological theory. Dawkins (1976), Wilson (1978), and Trivers (1981), for example, argue that anisogamy—the greater investment by women in the young beginning with the egg—may have favored *inter alia* a closer, more protective attachment to children on the part of women. The innate basis for this difference would presumably be reflected in the brain. Although the extent and larger significance remain unclear, sex differences in the human brain, anatomy, and physiology, are increasingly being reported (Springer and Deutsch 1989, ch. 7; see Benbow 1988).

Just as social scientists fail to speak in a single voice, neither do life scientists—especially those dealing with humans on a macro scale. Thus, Hrdy (1981), writing as a sociobiologist (and as a feminist), qualifies sociobiological theory and anticipates (though in the same way qualifies) Gilligan. Hrdy rejects the evolutionary picture of a more passive, cooperative female. In her survey of sex differences among primates generally, she draws a portrait of a competitive female, using her wiles actively to entice and manipulate males. Referring to the "myth" that stresses "women's natural innocence from lust from power" and "her cooperativeness," she concludes, "Such a female never evolved among the other primates" (p. 130). Rather, "competition among females remains a fact of primate existence" (cf. de Waal 1984, 1989).

Yet Hrdy (1981, p. 129) also refers to the "mistaken assumption that female competition will take the same form as competition between males." Indeed, Hrdy's woman concentrates her competitive strivings within the sexual arena; it is primarily a competition within her own sex for the most suitable mates that occupies her attention.

Somewhat ironically, both the Gilligan and Hrdy positions have been used by feminists: the Gilligan position supports an older nineteenth-century feminist view of women as innately possessing the "finer sensibility," and the Hrdy refutes the picture of a naturally docile female. But both views cannot equally hold, furthermore, both are in conflict with the contemporary feminist position of the fully competitive woman. This perspective, which sees women as the equal of men in every important respect and which views sex roles as entirely environmentally based, is summed up by Ethel Klein (1984, p. 3) as follows:

> For most of human history men's and women's different roles were accepted as biologically based and unalterable. This view is currently being challenged by the feminist movement, which asserts that women's roles are learned and socially determined. Whereas egalitarian roles were historically seen as radical, if not crazy and a threat to society, women today are willing to advocate sex equality, or the sharing of equal responsibilities within the home and marketplace.

Thus, we are confronted with three mutually incompatible theories of the human female role: (1) a cooperative and therefore less competitive role, both in relation to men and to other women (but noncompetitive would be too strong a term); (2) a sexually but not fully competitive role; (3) an egalitarian or fully competitive role. All three roles may apply, but the problem comes in their typological and overgeneralizing formulations. Even if there were a characteristic "female role," it would not necessarily apply to all women, and certainly not equally so. A populational approach is called for precisely because each individual is biologically unique; the inevitable existence of individual differences must always be kept in mind.

In order to see more clearly how these three views might fit the individual variation within the female population, we can view individual women as falling along two dimensions, varying in (1) degree of personal dominance from low to high; and (2) in space-time capacities from local to cosmopolitan, that is, precisely in two of the personal attributes—dominance and cognitive capacity—that one would normally associate with worldly success. Figure 7.1 locates the differing views of the female role within these dimensions.

As seen in Figure 7.1, the fully competitive woman, the feminist ideal, may exist, but she may be restricted to the ranks of the most dominant, cosmopolitan women (although, according to Gilligan, many sophisticated college women might not fit).

The fully competitive view may ironically tend to typologize the world of men, to identify it with power and dominance and, therefore, to overlook the great

Figure 7.1
Relation of Three Constructs of the Female Role to Dominance and to Local-Cosmopolitan Orientations

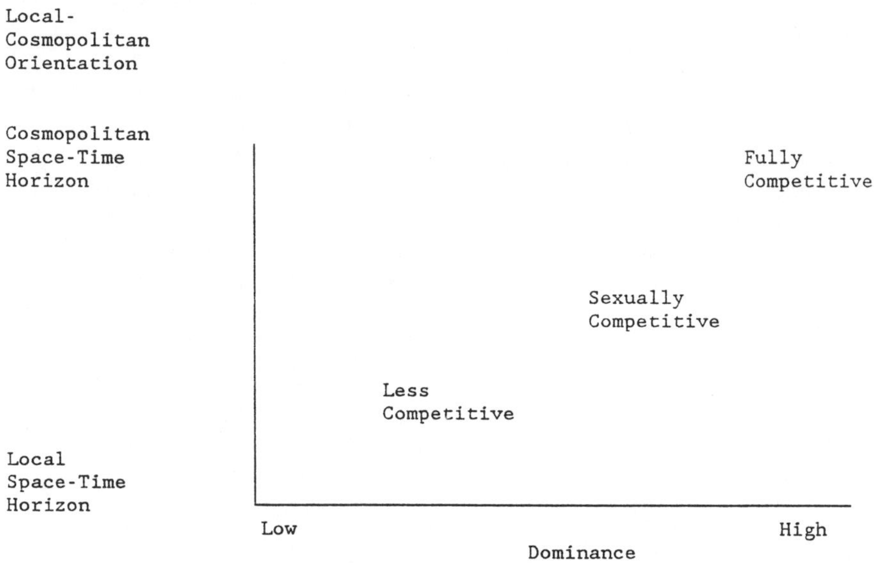

numbers of men who do not compete successfully within it. Keep in mind that men also vary in dominance and only a relatively few will appear in leadership roles.

The less competitive woman, not surprisingly, will be concentrated among the more local, less dominant women, while the sexually competitive woman falls somewhere between the two extremes. These three possible placements are examined in turn.

The fully competitive woman. "Modern societies as diverse as Great Britain, Israel, and India have had women national leaders," Masters (1989, p. 5) observes. Norway, Pakistan, and the Philippines can also be added to the list. Certainly a Margaret Thatcher or Golda Meir qualify as fully competitive players in the political arena, in every important respect, with no serious disadvantage of apparent difference according to sex. Therefore, even if the universality of the feminist ideal of the fully competitive woman remains problematic, its historic existence can be certified.

Masters (1989, p. 4) notes that a "painting by Berthe Morisot, a musical composition by Nadia Boulanger, a recording of a flute sonata played by Dorothy Doriot Dwyer or a scientific paper by Barbara McClintock do not seem to be specifically 'caused' by the artist's or scientist's XX chromosomal complement" (although culturally or intellectually based sex differences could creep in). Just as international communities in art and science may transcend ethnic, racial, and linguistic lines (even if artificial barriers based on these factors may continue to exist), so the same may be said of sex. Cosmopolitan attainment is based primarily on achieved, not ascribed status. Note that while such achievement has been identified with the world of the white male this formulation is misleading. Most white males may not qualify for entry, whereas, many nonwhites and females may. In any event, the more cosmopolitan and dominant the woman, the less likely she is to fit any theory that stresses innate and fixed differences by sex.

The sexually competitive woman. Nancy Reagan can be viewed as a prototype of the sexually competitive woman. Although a Smith College graduate and obviously intelligent, she does not fit the feminist model of the fully competitive female. And yet she surely cannot be seen as a passive, compliant female.

The paradox of Nancy Reagan and thus of the sexually competitive woman is captured by Kellerman (1981, p. 243):

> Perhaps the most economical way of viewing Nancy Reagan is to see her as a paradox. In part, she is an anachronism who appears for all the world to be living in a time when men ran things and women were less than equal, when they existed for and through their husbands. But she is also strong, opinionated, active, and fully participating in her husband's political life. The woman who says forthrightly that "my life really began when I married my husband" is the same one who, when her husband was asked by a reporter why he did not support the Equal Rights Amendment, stepped firmly between her husband and the microphone and insisted that she be allowed to "speak to that."

Regarding the first facet of this paradox, Gilligan (1982, p. 128) cites Elizabeth Cady Stanton as writing that "the thing which most retards and militates against women's self-development is self-sacrifice." By this standard, Nancy Reagan is no feminist. She willingly and without any regret gave up a Hollywood career as an actress to marry Ronald Reagan and thereafter devoted herself to advancing his career. By all accounts, including her own, she identified completely with her husband. Toward the end of the Reagan presidency, in an interview with Betty Cuniberti (*Philadelphia Inquirer*, p. 12, November 13, 1988), she said that "a first lady is a wife, first of all . . . that comes first for me . . . and I certainly felt that for Ronnie's interest." According to Mayer and McManus (1988, p. 361), "in an administration in which the ethic had become, as one aide to Reagan later conceded, 'every man for himself,' she was almost alone among those close to the president in putting his survival above everything else" (Mayer and McManus 1988, p. 361). Donald Regan (1988, p. 289) writes that as chief of staff he "tried to remember—though sometimes with difficulty—Mrs. Reagan's actions were motivated by a genuine belief that she was the best judge of her husband's interests."

The phrase "with difficulty" suggests the other—the competitive, assertive—side of Nancy Reagan. This side is first glimpsed in her education at Smith and her initial pursuit of an acting career. But the "sexually competitive" aspect of her nature dominates in her successful pursuit of Ronald Reagan. She acknowledges that she contrived her first encounter with Reagan and that her leaving acting after marriage "was of my choice" (Barrett 1984, p. 469). Note that in its first phase, such sexual competition may be largely intrasexual, that is women competing among themselves for the most suitable mates. In the second phase, however, at least for Nancy Reagan, such sexual competition may consist of thwarting off perceived threats to her perceived interests—in Reagan's case, her husband in particular, perhaps for most other women, the family more inclusively.

Nancy Reagan was no shrinking violet in defending her interests. She was instrumental during the 1980 campaign in having Reagan's first campaign director, John Sears, replaced by William Casey, and according to Woodward (1987, p. 138), Casey thereupon "had to assess the power centers around Reagan. The looks, the voices, the glances, the subtle deference told one story: Nancy." This was a lesson later learned the hard way by Donald Regan. Mayer and McManus (1988, p. 379) refer to Regan's "fatal miscalculation"—as the underestimation of the importance of Nancy Reagan.

The cooperative or less competitive woman. This woman comes closest to fulfilling the traditionalist submissive role. Like her male counterpart, she is less dominant and more localist in orientation. She has thus less cognitive as well as affective means to control her own life as well as the lives of others. The one conspicuous exception may reside in her own children, and it is here too, in her devotion to her family, that she may differ most markedly from her mate.

Why Locals May Be Characterized by More Sharply Drawn and Segregated Sex Roles Than Cosmopolitans

Is it true that more egalitarian and hence similar sex roles are more likely to characterize cosmopolitan populations? If true, then the reverse would follow: More unequal or more clearly distinct sex roles would characterize more localistic populations. Let us examine why such relationships may hold.

First, the general capacity to formulate an abstract space-time horizon must be placed within a broad ethological and developmental context. It seems clear that it is such a human capacity that permitted the development of a culture and the transcendence of immediate environmental constraints and of fixed genetically based responses (cf. Etkin 1985). The same relationship holds for the maturational process: the human individual attains a corresponding freedom of thought and action as he or she matures. Again, the reverse relationship should hold, namely, the less capable the individual or species in generalizing over time and space, the more tied either will be to environmentally and genetically based constraints.

A key question now remains: Do human adults vary significantly in cognitive capacity? Do all individuals effectively enter Piaget's fourth and final intellectual stage, that of formal operations? According to Kohlberg and Gilligan (1971), not necessarily. They report studies suggesting that "almost 50 percent of American adults never reach adolescence in the cognitive sense" with the result that "a large proportion of Americans never develop the capacity for abstract thought." Indeed, "in simpler cultures—for example, villages in Turkey—full formal operations never seem to be reached at all (though it is reached by urbanized educated Turks)" (Kohlberg and Gilligan 1971, pp. 1065–66).

We assume here a rough correspondence between Piagetian abstract thought and the capacity to formulate an abstract space-time horizon. Indeed, the defining characteristic of Piaget's final stage, the capacity to formulate scientific hypotheses, presupposes precisely the ability to transcend one's immediate place in space and time.

Second, we are willing to entertain the controversial proposition that *both* heredity and environment underlay individual variability. Further, such variability is laden with significance; nonetheless such variability often tends to get minimized or overlooked.

One important way in which cognitive differences may be overlooked is by overgeneralizing on the basis of studies conducted with small elite groups—just as Piaget may have overgeneralized on the basis of studies involving unusually bright Swiss school children. The same problem may arise in connection with studies conducted with college students, especially at select institutions. To the extent that this may be the case with respect to sex differences, such studies may *understate* such differences, that is to say, any biosocial differences that exist within the select group are likely to exist in magnified (if also qualified) form within the more general population. It is at least suggestive here that Maccoby (1965, p. 310) reports a 1936 study by Terman and Miles in which it was found that the higher the IQ of both men and women the more likely both sexes would register so-called masculine responses on various personality test items. The relevant point for us is that relatively less personality differences by sex characterized high-IQ individuals.

Gilligan's work (1982, pp. 2, 25) on sex differences may reflect this problem. At each age group, she clearly relies on the most intellectually advanced members for each sex, culminating in elite college students (Radcliff, Harvard). Thus the descriptions of males arriving at abstract understandings of ethical problems in a "postconventional" sense may not at all characterize most male adults, whose ethical behavior remains rooted within a more intellectually concrete, conventional mentality. It may mean, further, that the more mediatory, interpersonally sensitive responses of the young women—while in one sense more concrete but at the same time unusually subtle and sophisticated in nuance, might become more stereotyped and rigid for a less select population.

In order to elaborate these contentions, we will use the observations of Gans in *The Urban Villagers* (1965), the study of a close-knit working class neighborhood of Italian-Americans in Boston in the late 1950s and of his generalizations and comparisons to the broader American society. This analysis, however, remains at best suggestive, for there exists no research, at present, that directly tests the propositions just set forth.

Gans describes the life of the West Ender as being spent in and influenced by peer groups. Accordingly, "boys and girls form cliques or groups" which are ''sexually segregated'' and which maintain their hold ''until late adolescence or early adulthood,'' being ''broken briefly at marriage'' (p. 38). The ensuing marital relationship continues, according to Gans (p. 50; emphasis in original), to be ''*segregated*''; Gans cites a passage from Bott's (1971) *Family and Social Network* pertaining to English working class families as applying to West Enders as well: ''Husband and wife have a clear differentiation of tasks and a considerable number of separate interests and activities. They have a clearly defined division of labor into male tasks and female tasks. They expect to have different leisure pursuits, and the husband has his friends . . . the wife hers'' (cf. Komarovsky 1967).

In contrasting these sexual relationships to other social classes, Gans (p. 246) states that ''in the lower class, the segregation of the sexes—only partial in the working class—is complete.'' As children reach school age, Gans (p. 30) elaborates: ''The boys are allowed to roam the streets and to look for childhood adventures, more so in lower class than in working-class families. Girls, however, are expected to stay home.''

In the middle class, by contrast, marital relationships are ''joint'' (Gans, p. 30). Gans (p. 47) continues: ''Whereas the middle-class family is a companionship unit in which individuals exist most intensely in their relationships to each other, the upper-middle-class family is a companionship unit in which individuals seeking to maximize their own development as persons come together on the basis of common interests.'' Gans (p. 248) makes clear that in the latter ''professional work satisfaction is a focal concern'' not only for the man, ''but often for the woman as well.''

More broadly, Gans (p. 88) distinguishes class subcultures on the basis of whether their members are ''person-oriented'' or ''object-oriented,'' middle and upper classes being characterized by the latter. The object-oriented individual is motivated by ulterior aspirations: the person-oriented individual is much more guided in behavior by the effort ''to be liked and noticed by members of a group whom one likes and notices in turn'' (Gans, p. 90). This distinction, Gans (p. 91, ft. 12) informs us, is ''inspired most directly'' by Merton's distinction between locals and cosmopolitans. It is not surprising, therefore, that our local-cosmopolitan dimension dovetails with Gans's typology of class subcultures. (In her revised *Family and Social Network* [1971], Bott also explores the possible relationship between local-cosmopolitan orientations, the social class background of families, and their differing interaction.)

The one clear difference between our analysis and that of Gans (and Bott and Merton) resides in our assumption of an interactional as opposed to an environmentalist basis for the behavior described. At times, Gans (p. 254) implies the presence of an underlying genetic diversity, as when he notes that ''every population—even a low-income one—produces a number of young people with special gifts or talents,'' the West End being no exception. Such young people, however, usually find themselves ostracized by their peers, resulting in frustrated talent. Those who are ''especially talented and strongly motivated'' will, in the face of group ostracism, attempt to move out and upward in the outside world. They are ''object-oriented'' candidates for membership in the middle class.

Most West End children, however, are person-oriented and, as a result, do poorly within the object-oriented world of the school. Academic achievement, Gans (p. 133) observes, "depends largely on the ability to absorb and manipulate concepts, to handle the reasoning processes embedded in the lesson and the text, and to concentrate on these methods to the exclusion of other concerns." West End children "are adept at none of them." What is the cause? "Certainly it is not a lack of native or acquired intelligence." (From an environmentalist or utopian egalitarian perspective, there is either *no* such thing as "native" intelligence, or if there is, *no one* is ever deficient in it.)

If we insist on an underlying genetic dimension to individual differences in metaintelligence, in the capacity for extending space-time horizons, we must be mindful of individual differences. A populational approach will fully acknowledge the wide range of individual variation in ability that will likely characterize any population, just as Gans pointed out that a largely working class community, as in the West End, will produce children of unusual talent. An interactional perspective, furthermore, will fully grant the possibility that any native ability can be stifled because of unfavorable environmental conditions. With these caveats in mind, note that individual variations in capacity for space-time horizons account for at least some of the variation in sex roles tied to Gans's social class subculture typologies (n.b. that they *are* typological in character). The strict sex segregation of the lower or working class population seems a product of a mentality that is linked closely to the immediate social environment in the form of the peer group. For males or females, this is the sexually segregated clique. Note too how E. O. Wilson's "hypertrophy," that is, the cultural magnification of preexisting tendencies, might apply in such populations. For people whose conceptual range is limited might particularly be apt to stereotype and magnify differences that initially had some biological basis.

On the other hand, the individualistic, more egalitarian, and much less sexually segregated world of the upper middle class can be seen in part as a result of people coming together who share a cosmopolitan mentality in which self-realizing aspirations are pursued that may require either (or both) a long-term commitment or extensive travel. Table 7.1 links the relationships between sex roles, class subcultures, and local-cosmopolitan orientations.

At this point, additional caveats are in order. Suppose the lower and working class or localistic subculture produces a more sexually segregated society, together with greater numbers of less competitive women. The result would be, in feminist terms certainly, a sexist, male chauvinist society. But because of genetic diversity, generalizations are always risky. Even if more cosmopolitan abstract capacities and dominance may be loosely related (probably centered neurologically in the frontal lobes), they are by no means isomorphic. In short, localistic females might be quite dominant and more so, in some instances, than their mates. Even if they prove to be exceptions, they are exceptions that must be taken into account.

Now two further points concerning these more sexually segregated subcultures: First, the specific content of the sexual roles may depend heavily on the specific

Table 7.1
Relationship between Sex Roles, Class Subculture, and Local-Cosmopolitan Orientations

Class Subculture	Local-Cosmopolitan Orientation	Sex Roles
Upper Middle	Cosmopolitan	Relatively egalitarian and undifferentiated
Middle		
Working		
Lower	Local	Strictly segregated, differentiated, and likely unequal

historical circumstances because the predominance of the peer group translates as the predominance of the peer group as it *currently* exists, and its character must necessarily undergo some change with changing circumstances. This point leads to the second by raising the question: Are not derivations based on 1950s research necessarily dated? Contrary to feminist hopes, such has not proved to be the case regarding the basic outline of the relationships presented, as perhaps the subsequent story of the defeat of ERA may indicate.

If the example of Nancy Reagan as a sexually competitive woman is a valid one, it points to the significant presence of sex differences in both thought and action even in the upper middle class. Nancy Reagan was, after all, the stepdaughter of a wealthy neurosurgeon and a Smith alumnus who initially pursued an acting career with some success. Nonetheless, her example indicates clearly that the local-cosmopolitan dimension may have to be qualified in order to account for possible differences by sex.

The first signs of conflict between the first lady and Donald Regan occurred following the president's surgery in 1985. "Her concern for Reagan's health was all-consuming" (Mayer and McManus 1988, p. 113) and she bridled at the chief of staff's efforts to control fully the president's access and schedule. Regan's response was reported to be, "I've never let any broad push me around before, and I'm not about to start now" (Mayer and McManus 1988, p. 114). Regan (1988, p. 258) wrote later that "Mrs. Reagan regarded herself as the President's alter ego not only in the conjugal but also in the political and official dimensions, as if the office that had been bestowed upon her husband by its people somehow fell into the criteria of worldly goods covered by the marriage vows."

Here we have an apt suggestion of the sexual dimension underlying the territorial conflict between the first lady and the chief of staff. Regan's "territory" (based on a broad space-time horizon) was the entire executive branch of the federal government (if not more), including control over access to the president and over his schedule. The first lady's "territory" was literally her husband, and however much dominance and cosmopolitan-based skill she brought to bear to protect her turf, that turf was (initially at least) more narrowly delimited in scope. Her constant protectiveness of the president proved a critical factor, according to another account (Mayer and McManus 1988, p. 22), in inducing the first Reagan chief of staff, James Baker, to relinquish his post to Regan, and then in removing Regan several years later. But for all the momentous political consequences of her actions, they appear to be defined by a role that can best be described as sexually rather than fully competitive.

If the action of a cosmopolitan first lady can be influenced by her sex, presumably such will be all the more the case for more localistic-oriented women. Gans (1965, pp. 47–48) observes because "the Italian culture is not a drinking one," much of their sexual segregation occurs within the home itself, with one room reserved for the men, another for the women. But he also notes, "In many working-class cultures, the man is away from the house even after work, taking his leisure in the corner taverns that function as men's clubs." Here again we have an indication of how the local-cosmopolitan dimension may have to be qualified by differences in sex. The localistic world of men and women may in particular radically differ, in some cases with significant political ramifications. Take, for example, the rise of the temperance movement in late nineteenth-century America: "At the time it was recognized, as it has been since, that the impetus behind women's opposition to drink was concern for the moral well-being of the home. . . . The saloon was a man's world, one in direct competition with the home, the proper sanctuary of the family, that is, all the family, not only mother and children but the father as well" (Degler 1980, p. 16).

SELF-SELECTION AND MOTHERHOOD: MATERNAL INSTINCT AND MATERNAL AUTISM

If there is a biological basis for mothering, this activity should receive greater centrality within the lives of locals in contrast to cosmopolitans for two reasons: (1) a lesser space-time horizon implies being tied more closely to immediate environmental contingencies (e.g., a peer group) as well as genetic predispositions; and (2) a broader space-time horizon, on the other hand, permits, if not implies, the possibility of interests and concerns that may compete with or transcend any such predispositions.

Let us explore this question and its larger sociopolitical ramifications. Many feminists take pains to dissociate maternal behavior from any biological foundation. Thus, in grudgingly conceding a biological tie to an infant through pregnancy and birth, Klein (1984, p. 60) argues that the "radical transformation of infant feeding practices" has "eliminated the biological basis for women's responsibility for infant care." A once female responsibility now becomes potentially male as well as societal. Such a position has invited counterattack from Fraiberg in *Every Child's*

Birthright: In Defense of Mothering (1977), which, based on the pioneering work of Bowlby, Spitz, and Harlow, argues for the vitality of an innate mother-infant bond.

An interactional approach would prima facie reject the idea of a fixed, uniform maternal instinct, but neither would it accept the universal maternal autism that more extreme feminists espouse. The term *autism* is indeed an oxymoron, in suggesting that it is a matter of indifference to a mother who cares for her child. Autistic individuals are, of course, so self-centered that normal social interaction is inhibited. As Melone and Lettick (1981, p. 172) observe, ''The autistic person's inability to understand subtleties, nuances, the give-and-take of flirtation, dating, and love will then preclude any realistic relationship with a normal partner.'' Severely handicapped autistic persons do not marry and have children. Finally, as Schopler (1983, p. 6) states, ''increasing evidence shows that a variety of biological factors may produce the syndrome of autism'' (cf. Springer and Deutsch 1989, pp. 274–77). In other words, the counterexample of autism strongly suggests that all normal human social interaction has a biological basis. The mother-child relationship would not likely be an exception.

A typological view that rigidly isolates severely autistic and entirely ''normal'' humans, with little variation, may also be in error. A populational approach suggests, rather, that even if sociality is a normal human quality, it may vary widely on an individual basis in a population. In particular, a maternal predisposition, in the form of a desire to have and to raise children, may not exist uniformly. With married couples unable to have children, it is not usual for the woman in particular to desperately want to ''correct'' the situation. Conversely, a certain proportion of married couples are characterized by ''voluntary childlessness'' (cf. Veevers 1974). Veevers (1974, p. 308) implies that this reality undermines a biological explanation for parental motivation. However, the fact that a few autistic humans are born incapable of forming normal social relationships does not mean that a biological basis for such sociality is therefore absent for all other humans. In the same way, it is possible that a wide variation in parental predisposition exists. Obviously, it will be affected by environmental circumstances, but can biologically based individual differences be entirely ignored in comparing two sets of childless couples, those voluntarily so and those desperate to change their condition? It is instructive to note, following Wolf's case study of a lesbian community (1979), the immense variation among lesbians toward mothering—from the separated mothers determined to keep their children to extreme feminists who eschewed biological motherhood and children. Finally, in Goodall (1986, pp. 101–3, 379–80), we can see variation in apparent maternal attachments to their offspring in chimpanzee females. Granted that some of this variation has an environmental basis—especially in the case of mothers who had been orphaned when young, but even this fact points to the biological centrality of the mother-child bond.

If there is a biological basis for mothering, does it also exist for fathering? And if it does, is it equally strong? Maccoby (1966, pp. 330–32) surveyed fifty-two studies of ''nurturance and affiliation,'' which indicated that in forty-five of these girls and women scored higher in these qualities than boys and men (in five there were no differences; in two the males rated higher). In a recent nationwide survey of American mothers (Genevie and Margolies 1987, p. 21), overwhelmingly fathers were reported to play a secondary role in the caring of their children, even when both parents had outside employment.

In any case, this chapter makes three critical assumptions concerning a parental predisposition:

1. It is genetically and environmentally influenced.
2. The strength of its genetic influence may vary widely in any population.
3. Women are, as a whole, likely to possess a stronger biological predisposition for parenting than men.

If nothing else, these propositions appear more plausible than any set of counterassumptions. Following Masters's "model of gender difference" (1989, p. 5), these assumptions are illustrated in Figure 7.2.

Returning to our emphasis on local-cosmopolitan orientations, parental disposition is not likely, any more than dominance, to be isomorphic with cognitive capacity. Table 7.2 indicates how varying parental dispositions in women may interrelate with varying local-cosmopolitan orientations to yield different self-selecting tendencies and potentially important sociopolitical consequences.

The three categories marked in Table 7.2 are appropriate for discussion. The first, labeled "career first" women, is characterized by cosmopolitan space-time horizons and a low level of parental predisposition. Studies of voluntary childless couples (cf. Popenoe 1974; Gustavus and Henley 1974) indicate that such couples are typically of high socioeconomic status, with both husband and wife much better educated than the normal American couple. These couples are also more likely to be nonreligious. A frequent motivation involved cases "in which the wife gave up maternity to work, not because she needed the money but because she preferred

Figure 7.2
Sex Differences in Parental Predisposition

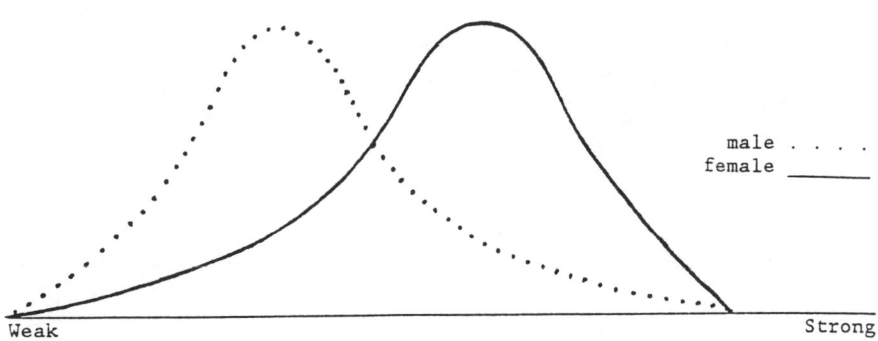

Table 7.2
Key Relationships Involving Parental Predisposition in Women and Local-Cosmopolitan Orientation

Local Cosmopolitan Orientation		
Cosmopolitan career Space-time Horizon	1. Career first (Fully competitive woman)	2. Torn between career and family (Less than fully competitive)
		3. Family first (sexually competitive)
Local space-time Horizon		
	Weak	Strong
	Parental Predisposition	

the outside occupation, and did not want to interrupt it'' (Popenoe 1974, p. 780). It is not hard to see a cosmopolitan self-selection process at work here. (It should be made clear that ''career'' as used here denotes a long-term commitment to a vocation, whereas, many studies report, most jobs taken by women in the work force are short-term.)

Another, if less often explicitly stated reason for voluntary childlessness, is a ''plain dislike for children.'' One couple, for example, said: ''We are unable to tolerate the presence of children for any appreciable length of time and cannot imagine any circumstances in which either of us would ever want children of our own'' (Gustavus and Henley 1974, p. 242). If we include in this category the single women who also qualify, it would appear that this group would come close to fulfilling the feminist ideal of the fully competitive woman and would include some of the most ardent feminists (cf. Cohen 1988).

The second category involves women who have cosmopolitan talents and aspirations, on the one hand, and an intense desire for children, on the other. They may, thus, be torn between career and family. Sophia Loren (cited in Sorel 1984, p. 317), who at twenty-nine ''had become obsessive about turning thirty without having produced a child,'' wrote that ''my urge to be pregnant intensified. The actress part of me was happy and fulfilled, but the other half of me, the Neapolitan, childbearing, Mother Earth half of me, was fiercely unsatisfied.'' Note that such conflicting pressures are probably less likely to occur for men.

This intense desire for a family with children by such women may make them as much sexually as fully competitive, to the point of sacrificing their career in more extreme instances (e.g., Nancy Reagan).

Cosell (1985, pp. 57, 104), in her case study of successful American career women, complains that "women were raised, and probably still are, with too much emphasis on marriage, or acquiring an identity and self-esteem through a man." Commenting that a "variety of successful professional women in the early 1980s, women who are supposed to be so different and changed from their mothers and perhaps their grandmothers, sound remarkably like those women when they talk about themselves," Cosell, in a chapter called "Success Schizophrenia," remarks: "Schizophrenia. Divided self. Torn in half. Not exactly what the popular press conjured up when it investigates the status of successful women in America. But I hear these words, these descriptions, over and over again" (p. 71). It is not hard to see some support for the corporate "mommy track" among such women. Their feminism is likely to be tempered by the pull of their families.

A good example might be Sylvia Ann Hewlett, whose book *A Lesser Life* (1986) reflects her being one of those "mortals" who are "trapped between the demands of the earth mothers and the hard-nosed careerists" (p. 16) and who alludes to her four children as prompting "the questions that lie behind this book." Early in her academic career, she experienced frustration as Barnard College refused (at that time) to grant maternal leave and, in the end, denied her tenure. Hewlett (pp. 197–98) writes that she had given ERA "knee-jerk support" initially as an "East Coast liberal," but that now she "could no longer be counted on to vote for the ERA."

The cross-pressures can also work the other way to temper an anti-ERA or pro-life viewpoint. Luker (1984, p. 145) quotes one prolife woman in California who had become rather inactive: "I guess I'm pro-life in a very large sense of the word, and I have found that a lot of people involved in the Support Life League are very narrow in their anti-abortion. . . . I'm not that kind of pro-life. I'm pro-life in that I'm interested in things like world hunger, and I'm against capital punishment, and I'm against war, and a lot of things that are broad areas of being pro-life." It seems plausible that this prolifer has a more cosmopolitan sensibility than most of the others.

The presence of a relatively stronger parental predisposition in cosmopolitan women in contrast to such men fits the sociobiological (if too typological) picture of women being especially attached to their offspring and to their future safety and well-being. The result may be contained in the observation of two feminists (Seager and Olson 1986, p. 39): "Peace has always been a women's issue: women have been in the lead of anti-war and anti-militarist movements for the last century."

The third category to be considered here involves those women who feel a strong urge toward marriage, family, and children but who are neither cosmopolitan nor narrowly local in orientation. Just what their orientation might be will be explored in the section to follow. The main point here is that such women are, on the one hand, not as likely to be so strongly motivated by any cosmopolitan career aspirations so as to be distracted from devoting themselves wholeheartedly to their families. Yet, on the other hand, neither are they so lacking in cognitive skills as to be unable staunchly to defend what they perceive as their vital interests. Those interests are no doubt initially tied to the major thrust of their self-selecting efforts: the maintenance, growth, and defense of their families.

LOCALS, COSMOPOLITANS, AND THE END OF ERA

As a prelude to a discussion of the Equal Rights Amendment, it is necessary to introduce, as shown in Table 7.3, an expanded outline of differing local-cosmopolitan orientations, their possible political ramifications, and the special emphasis that women may invest in them.

As chapter 4 elaborates, Merton's original local-cosmopolitan distinction (Merton 1957) is extended into a broader continuum marked by six main categories. The cosmopolitan, likely to constitute a small minority in any large society, is broken down into three separate categories, depending upon where the primary cognitive focus lies—the international, national, or sectional level. The local is in turn subdivided by community and neighborhood. In other words, the relatively small number of community activists who, according to Merton (1957), ran Rovere, New Jersey are distinguished from the vast noncosmopolitan majority who are more closely

Table 7.3
Relationship of Local-Cosmopolitan Orientations, Political Activity, and Likely Female Emphasis

Local-Cosmopolitan Orientations	Political Activity	Likely Female Emphasis
International Cosmopolitan National Cosmopolitan	Initiatory (and extensive) political activity at national and international levels	Broad concern with "women's" issues, etc. sexual equality
Sectional Cosmopolitan Community Local	More reactive political activity at sectional and local levels	Primary focus on family and those political issues perceived as directly affecting it
Neighborhood Local	Minimal political activity; reactive to personal "bread and butter" issues	Largely non-political focus on family
Social Isolate	apolitical	

tied to their immediate neighborhood (as was graphically the case with Gans's West Ender).

The local, whose vision is more narrowly circumscribed, is more likely to be tied to more traditional ways of thinking and living; thus he or she may well react negatively to cosmopolitan initiatives. Such reactive politics, from a cosmopolitan perspective, invite being stamped as "reactionary." Community locals may possess highly effective political skills. Their space-time horizon remains sufficiently broad to permit the formulation of appropriate political tactics—tactics that may frequently come to be viewed as shrewd. Indeed, being less prone to focus on "remote" questions and being more in tune with the personal, instead of the abstract, side of the question, the community local may be especially effective in exerting political influence on a grass-roots level. The community local will further be aided by relatively greater numbers (as against cosmopolitans), and it is likely that the bulk of political officials on the state and local level may come from their ranks. As Burns (1949) pointed out some time ago, the local congressman is often the "hometown boy" who made good.

The possible political clout of the community local should therefore not be discounted. This observation applies with equal force to women who hold this role, even if the local-cosmopolitan dimension may have to be qualified in order to apply fully to them. Merton's "community locals" apparently included women, but he did not explore any differences by sex, except to note that women were much more likely to have been influenced by other women (Merton 1957, p. 467). It is perhaps true that, more so than for men, continuous female political activity may be restricted to a small cosmopolitan category. The community local woman may be more sexually than fully competitive, her focus more centered upon her family. Nonetheless she will retain the ability to intervene directly in the political process (especially at the local level) whenever she feels her basic interests threatened. Just as in the case of Nancy Reagan, the unit of (defended) territory may be more narrowly and more personally defined for the woman, but its political defense may nevertheless be effective.

Neighborhood locals—both male and female—may be less effective politically. Tied to the immediate and concrete, as with Gans's West Ender (1965), they are likely to be preoccupied with the everyday personal concerns that Stouffer (1958) documented so well for most Americans a generation ago. It is directly perceived threats in these areas—often the bread-and-butter, pocketbook issues—that will typically generate whatever limited political activity may ensue.

THE FIGHT OVER ERA

The First Phase: Cosmopolitan Sponsorship and National Triumph

The highly educated background of the active supporters of the contemporary women's movement and of the Equal Rights Amendment (ERA) has been well established (cf. Boles 1979, p. 47; Mansbridge 1986, pp. 226–27). The likely cosmopolitan character of these supporters is perhaps epitomized by Betty Friedan, the founder of the National Organization of Women (NOW) and one of the leading proponents of the ERA. Born and raised in Peoria, Illinois, she attended Smith College where she graduated summa cum laude with a college official stating that she "has the most outstanding record of any student ever matriculated at Smith" (cited in Cohen 1988, p. 62). There seems clearly to be a process of cosmopolitan self-selection at work in her moving out West to study psychology at the University of California at Berkeley upon graduation from Smith, and in her eventual (if temporary) sojourn to Greenwich Village (instead of returning to Peoria). For the record, though, her turning down, while at Berkeley, an attractive, prestigious fellowship in order to pursue a romantic interest indicates the presence of a sexually-competitive streak in Friedan. It is also of note that she never shared the anti-male, anti-family animus of the more radical feminists.

Friedan's career as a writer was to involve her in reporting the results of a survey of Smith College alumni in a book published in 1963 called *The Feminine Mystique*. Basically she reported that for this elite group the process of cosmopolitan self-selection was being frustrated all too often by the continuing status of the majority of the women in the sample as full-time mothers and homemakers. Friedan (1974 [10th Anniversary edition], p. 360) refers to these "200 Smith graduates" as having "their counterparts in women all over the country, women of intelligence and ability, fighting their way out of the housewife trap." She (1974, p. 366) urges educators to "expect the girl as well as the boy to take some field seriously enough to pursue it for life," maintaining (1974, p. 337) that, "Only by such a personal commitment to the future can American women break out of the housewife trap." Just as I have asserted elsewhere (E. White 1989), the more intellectually abstract the individual, the more likely that he or she may feel trapped by circumstances. Basically, the same roles of mother and housewife that one woman might happily aspire to and permanently assume might alienate, or at least not in and of themselves fulfill, another.

Thus Chafe (1972, p. 231) argues that "Friedan and the other feminists concentrated their attention on suburban, college graduates," overlooking that "millions of lower-class women, on the other hand, lacked both the sophistication and experience to envision the possibility of an alternative life style." (I would of course frame this critique in populational local-cosmopolitan terms rather than with typological social class constructs.) It is perhaps suggestive that in 1976 fully three-fourths of American housewives (76 percent) with a grade school education reported that they "enjoyed" housework, as against 38 percent of those who attended college (cited in Mansbridge 1986, p. 106). In any case, it is the subjectively trapped cosmopolitan woman who would, I submit, especially advocate sexual equality.

Friedan's cosmopolitan orientation also comes through clearly in her account (Friedan 1985, p. 191), of why she preferred writing a magazine column for a mainstream American woman's magazine (*McCalls*) instead of a feminist journal.

"I would be bored to write only about women's liberation. Women now are as involved with the life-and-death questions of war and peace, the crisis of the planet and our cities, the unknowns of theology and art and space, as they are, still, with food and children and home."

It is perhaps not out of place to note here the cosmopolitan orientation of most women academics, which would tend to make them natural supporters of the ERA. Indeed, academic scholarship concerning the ERA is overwhelmingly feminist in nature, as typified by the title of the study, "Why We Lost the ERA." The author (Mansbridge 1986, p. 8) could not have justified her choice in more cosmopolitan terms:

> Who, then, is the "we" of the title? First of all, it designates those who supported the ERA. . . . But the "we" also represents . . . those who opposed the ERA and those who did not care. Almost everyone involved in the ERA struggle was primarily concerned with the good of the country rather than personal self-interest.

She goes on to conclude that since she believes the ERA "would have done more good than harm for the nation as a whole over the long run," that thus even the opponents were "really" the "losers" as well. Again, note the quite explicit cosmopolitan space-time horizon. A cosmopolitan orientation also suffused NOW, at least as translated in national as opposed to local politics. Mansbridge (1986, p. 171) observes NOW was "a national organization, resting even more on its direct-mail membership than on its constituent chapters."

> Once NOW became visibly involved in the national ERA struggle, its direct-mail membership began to increase dramatically. This meant that the organization naturally leaned toward . . . television ads and massive demonstrations that would get national coverage, rather than local organizing in districts with wavering legislatures, many of which did not even have a NOW chapter.

It is not incidental to note that the National Organization of Women was indeed national in origin, beginning in the nation's capital. Friedan (1985, p. 77) credits Catherine East of the Women's Bureau of the Labor Department as the woman who "cajoled me into the act of starting NOW" in 1966 and whom she further identifies as "the pivot of the feminist underground in Washington spreading from government agencies to Capitol Hill."

This Washington-centered cosmopolitan network paid rich dividends at the national level in the fight for the ERA following the passage of the Civil Rights Law in 1964. With the outlawing of discrimination by sex in employment and with subsequent court decisions negating protective women's legislation (that some feminists and labor unions had favored as against the sexual equality mandated by the ERA), the road was finally cleared by 1971 for passage of the ERA by Congress. As Hewlett (1986) relates, "ERA opponents were hard to come by. Hubert Humphrey was for it, but so was George Wallace. Bella Abzug rooted for it, but so did Spiro Agnew.'' The ERA passed the House in October, 1971 334–23 and the Senate approved it 84–8 the following March. Within three months twenty states proceeded to ratify the ERA and by early 1973 thirty states had approved. With six years to spare, only eight more states would be needed to make the ERA a part of the Constitution.

The Second Phase: Local Backlash and Defeat in the States

''While the national offices of the various pro-ERA organizations could relatively easily coordinate their Washington activities to get the ERA through Congress,'' Mansbridge (1986, p. 13) observed, ''They were slow in organizing coalitions in the states.'' Of Illinois, the author's own state, Mansbridge (1976, p. 169) writes:

> The typical NOW member had not been born in Illinois and did not necessarily expect to spend her life there. In my local NOW chapter, many of the members were recent migrants to the city. . . . One NOW leader summarized her constituency: ''We have a lot of people who know about . . . all the feminist issues—but they don't know a thing about street lights.'' While these NOW members might eventually settle in Illinois and become local political leaders, right now they were ''cosmopolitans'' who had joined the organization primarily to share their feminism in an atmosphere of mutual support.

The opponents of the ERA were more likely to know about, if not the state legislature, the neighborhood street lights. Indeed, Schubert (forthcoming) asserts that if the pro-ERA forces ''made the mistake of proceeding from the top down, establishing several national chapters in the states, which in turn promoted local chapters,'' the leading opponents gained from organizing ''from the grassroots'' and emphasizing ''local chapters rather than national organizations.'' Conover and Gray (1983, p. 75) refer to ''hundreds, if not thousands,'' of these ''grassroots groups'' that formed ''on an ad hoc basis.''

Now if the ERA proponents tended in fact to be cosmopolitans more adept in thinking and acting at a national level, were anti-ERA activists more localistic and hence likely to be better organized at a grassroots level? If we return to early 1973, we discover an event that inspired just such an anti-ERA opposition. It consisted initially of a backlash directed against the January 1973 *Roe v. Wade* decision of the Supreme Court which legalized abortion throughout the entire nation and removed all restrictions on abortion during the first trimester of a woman's pregnancy. The ERA did not directly touch on abortion and apparently it remained remote when initially passed in Congress to most of those who would become mobilized by the abortion decision. But there would be a tangible spillover effect. Mansbridge (1986, p. 13) puts it this way:

> Since NOW was not only the leading sponsor of the ERA but the leading defender of abortion on demand, conservative activists saw abortion and the ERA as two prongs of the ''libbers''' general strategy for undermining traditional American values. Unable to overturn the *Roe*

decision directly, many conservatives sought to turn the ERA into a referendum on that decision. To a significant degree, they succeeded.

Conover and Gray (1983, p. 185) assert that the "link between feminism and abortion" was "forged in the minds of the members of the New Right." Arrington and Kyle (1978, pp. 169–70), report that the anti-ERA activists in North Carolina were "largely opposed to abortion on demand, while most of the pro people favored it" and that "The women in each group were more unified on that question than the men." (Some of the anti-ERA men took a "perfectly consistent libertarian" or "anti-government" position, thus opposing the government regulating either abortion or business. This latter position, by the way, characterized and motivated the initial opposition to the ERA—in 1972, before *Roe*—by its most prominent and cosmopolitan foe, Phyllis Schlafly [see Conover and Gray 1983, p. 85], who has maintained a lower profile in pushing for government restrictions on abortion. Schlafly, a Harvard M.A. and author of nine books, is indeed a fully competitive woman, who has drawn grudging respect from feminists. Past NOW president Karen DeCrow said, "I admire her. . . . She's an extremely liberated woman" [cited in Klatch 1982, p. 143].)

Who, then, were the people who were mobilized to counterattack on the issue of abortion? I rely here on Luker's in-depth study of activists on both sides of the abortion question in California in the late 1970s (1984). Interestingly, no significant demographic difference emerged to distinguish the male activists on either side. However, since abortion (like the ERA) has been predominantly a women's issue, contested most strenuously by women, (women comprised over 80 percent of Luker's sample of activists on both sides) and since these women differed "diametrically" on social background variables in ways quite similar to the pro- and anti-ERA forces (in at least some cases the same women were confronting one another in both cases), I will, like Luker (1984, ch. 8), focus on the female activists.

Socioeconomic background. The pro-choice activists were better educated (92 percent completing college versus 70 percent of pro-life women). An "astounding 94 percent of all pro-choice women work, and over half of them have incomes in the top 10 percent of all working women in this country," while 63 percent of pro-life women, by contrast, do not work in the paid labor force, and almost all of those who do are unmarried" (Luker 1984, p. 195). While pro-choice activists were often employed in the major professions, as administrators, owners of small businesses, or executives in large businesses, the pro-life activists, by contrast, were mostly housewives, even the 6 percent with an advanced degree (compared with 18 percent for the pro-choice women, who were much more likely to be employed).

These findings are consistent with those relating to the ERA. Thus Arrington and Kyle (1978, pp. 172–74) also report no significant differences for the male activists on either side in North Carolina, but note that the anti-ERA women activists were "not as well educated," and were "overwhelmingly housewives" (while "almost all the pro-ERA women worked outside the house, and more than half had professional occupations"). Income differences, however, were not significant, although here Arrington and Kyle refer to total family income, whereas Luker (1984, p. 195) carefully distinguishes between family and personal income. In doing so, Luker reveals clearly that pro-life and pro-choice women differed markedly in the latter respect, as must have also been the case for the female ERA activists.

Marital status and family life. We have already observed that both pro-life and anti-ERA women were much more likely to be housewives than their opposing counterparts. Indeed, they were also more likely to be married, less likely to be divorced and separated, and much more likely to have a large family. At least this finding is clear for Luker's pro-life women (1984, p. 196), where the modal pro-life family had three children versus one for the modal pro-choice family. The number of children of activists on both sides of the ERA was not mentioned in the ERA studies I consulted; I can only refer here to a 1982 general survey of American public opinion on the ERA (Mansbridge 1986, p. 213), in which support for the ERA increased from 62 percent for those with 6 or more children steadily upward to 78 percent for those with no children (among those married or once-married).

Sexually or fully competitive women? Luker (1984, p. 197) describes the "average" pro-choice activist as a women who married at twenty-two, has one or two children, has "some graduate or professional training beyond the B.A., whose own father was a college graduate and whose husband is a professional man and whose joint incomes from regular employment exceeds $50,000 a year." In contrast, the "average" pro-life activist married at 17 and has "three children or more." Her father was a high school graduate and her husband "a small businessman or a lower-level white-collar worker." Although she has some college or a B.A., she is "not employed in the paid labor force," and their family income totals $30,000 a year.

Finally, as implicit in Luker's portrait for the pro-life women, the anti-ERA women were more likely to have improved their status by means of pursuing their formal education further than their fathers and thereby marrying upward. The old 1950s pre-feminist strategy of attending college in order to receive a M.R.S. may not be wholly extinct. Just as Nancy Reagan dropped her career when she wedded Ronald Reagan, both pro-life and anti-ERA activists seem to have ceased both their education and employment upon marriage.

Moreover, the pro-life women married well, at least in socioeconomic terms. Brady and Tepin (1976, p. 572) report that in their sample of Texas anti-ERA activists, "Eighty percent of the women with college educations are married to men of relatively high status (professional or business) or to ranchers." They (1976, p. 570) observe that over half the women have gone to college, an "unusually high" proportion for a sample of older Southern women, but that "despite this level of education, 70 percent of the sample are nonworking housewives." Nonetheless, family income (i.e., the husband's) is "high," especially for the husbands of the college educated women.

"Using education as a standard," Arrington and Kyle (1978, p. 678) write, "the women opposed to ERA were the most upwardly mobile group" (compared to both pro- and anti-ERA men and pro-ERA women), which is partly explained by the fact that "their fathers had the least education." Indeed, while "only 16 percent of the state's population over twenty-five years of age had any formal schooling beyond high school, 62 percent of the anti-ERA females have more than a high school diploma." However, 62 percent were also homemakers, as opposed to only 13 percent of the women supporting the ERA (Arrington and Kyle 1978, p. 673). Yet total family income did not differ significantly, with all four groups being "rich compared with the people in North Carolina as a whole." The sexually competitive woman who succeeds thereby perpetuates hypergamy.

By inferences based upon observed behavior patterns alone, differing patterns of self-selection appear to characterize the pro-life, anti-ERA activists, on the one hand, and the pro-choice, pro-ERA women, on the other. The latter fit the fully-competitive category described in previous sections, the former the sexually-competitive category. It now remains to illustrate this contention by reference to the differing outlooks of these women.

The differing world views. To use Luker's terms (1984, p. 159), each side of the abortion question has "an internally inherent and mutually shared view of the world that is tacit, never fully articulated, and, most importantly, completely at odds with the world view held by their opponents." The pro-life activists believe that men and women, being intrinsically different, are fully justified in adopting traditional sex roles that stress different responsibilities. In particular, women should be mothers and homemakers, nothing being more important than the bearing and rearing of children. Accordingly, sex exists only for purposes of procreation and abortion is under all circumstances (with the possible exception of when the life of the mother is at stake) an evil.

The pro-choice mentality stresses the substantial equality of the sexes. Therefore it advocates sexual equality (in effect that women should be fully competitive). Accordingly, the pursuit of a career, just as for males, becomes a desideratum, to the point of potentially taking precedence over marriage and children. Sex is considered good in itself, aside from procreation, and abortion is justified to remove an "unwanted" pregnancy. This very phrase is an anathema to pro-life activists.

These opposing world views seem equally to characterize the pro- and anti-ERA women. I have already referred to Betty Friedan's cosmopolitan, fully-competitive position. Sylvia Ann Hewlett (1986, p. 185, emphasis in the original) states that when Friedan wrote *The Feminine Mystique* she referred to the deep well of frustration held by homemakers "as the problem that has no name," adding, "One might say that *motherhood is the problem that modern feminists cannot face*." Such feminists, she writes (1986, p. 179), made "a gigantic mistake." "They assumed that modern women wanted nothing to do with children. As a result, they have consistently failed to incorporate the bearing and rearing of children into their vision of a liberated life."

Indeed, the feminist pro-ERA position was actively portrayed as anti-motherhood and anti-family by the anti-ERA activists. Boles (1979, p. 106), for example, refers to anti-ERA literature that charges that, "The Amendment would simply dynamite the national foundation of marriage and our traditional family life." Luker (1984, p. 206) cites one pro-life activist as opposing the ERA because we are "on a pedestal, why should we go down to being equal?" The sexually (as against fully) competitive woman openly covets "the pedestal" of a protected, supported marriage role within the traditional family.

The operation of self-selection: religion. Luker does not explore the underlying basis for the division between pro-life and pro-choice activists, but rather implicitly assumes an environmentalist position. But she provides enough material to allow for an alternative explanation, such as one calling for an explicitly biological dimension and the active operation of a process of self-selection. Luker (1984, p. 199, emphasis in original), for example, notes, "The values that lead pro-life and pro-choice women into different attitudes toward abortion are the same values that led them at an earlier time to adopt different lifestyles that supported a given view of abortion":

> For example, pro-life women have *always* valued family roles very highly and have arranged their lives accordingly. They did not acquire high level educational and occupational skills, for example, because they married, and they married because their values suggested that this choice would be the most satisfying life open to them. Similarly, pro-choice women postponed (or avoided) marriage and family roles because they chose to acquire the skills they needed to be successful in the larger world, having concluded that the role of wife and mother was too limited for them.

Let us take a closer look at the possible operation of self-selection in connection with the one factor that has perhaps been most clearly and strongly linked to both the abortion and ERA issues, that of religion. The "single most dramatic difference between the two groups," according to Luker (1984, pp. 196–97), is "the role that religion plays in their lives." That role is practically nonexistent for pro-choice, but overwhelmingly important for pro-life, women. Although 80 percent of pro-choice people never attend church, only 2 percent of the pro-life forces never do. Nearly 80 percent of the latter are Catholic.

The ERA studies also stress the role of religion, especially in connection with the anti-ERA women. Thus in Texas, fully 98 percent of anti-ERA women activists, Brady and Teden (1976) report, are members of a church; and 92 percent say that religion is "very important" to them, the most extreme of the possible responses. Two thirds are fundamentalist Protestant, compared to less than 8 percent in the state as a whole. Similarly, in North Carolina, according to Arrington and Kyle (1978, p. 674), the anti-ERA women in particular "overwhelmingly belonged to the major fundamentalist denominations of the state, while the other three groups (anti-ERA men and pro-ERA men and women) split mainly between non-fundamentalist churches and no affiliation at all," with the pro-ERA activists more likely to fit the latter category (cf. Matthews and DeHart 1990, pp. 124, 153, 175ff).

Where does self-selection come into the picture? As Luker noted, while almost 80 percent of the pro-life women were Catholic, only 58 percent were raised in that religion. And while absolutely none of the activist pro-choice women in this study were Catholic, 20 percent had been brought up as Catholics. Thus, as Luker (1984, p. 196; emphasis added) points out, ''almost 20 percent of the pro-life activists in this study are converts to Catholicism, people who have *actively chosen* to follow a given religious faith, in striking contrast to pro-choice people, who have *actively chosen* not to follow any.''

Since Luker provides no personal examples, I will go elsewhere for possible illustrations of the active self-selection process involved. A Catholic anti-ERA activist in Massachusetts, described in Klatch (1982), was brought up as a Catholic and attended an all girls Catholic college. ''Mary Donnelly'' went to work as an analytic chemist for two years before marrying. She gave up her job and eagerly anticipated having children, the first arriving within a year. ''I had five kids in six years, so I was very busy. It was marvelous. I loved it'' (Klatch 1982, p. 157).

Donnelly never left the church, but her faith was severely shaken once. ''I lost the faith for awhile. With my daughter's death. I felt, 'How could the Lord let this happen?' '' It is worth noting that if having children is the most central and satisfying event in the world for some women, then any threat to or interference with that activity may inspire a strong hostile reaction, against either human feminists or even the Lord himself, for a devout woman. Donnelly eventually returned to the faith. Although she had earlier also been upset with the church when it ''got into psychology and all of that—no longer taught right and wrong'' (Klatch 1982, p. 154), she was later reassured when she read the Papal Encyclical ''Humanae Vitae,'' which she says stressed ''responsible parenthood'' and renewed my faith in the infallibility of the Pope, particularly on moral issues (Klatch 1982, p. 661). Note here that Donnelly is not simply passively accepting whatever the Church and the Pope set down, but rather that which already fits into her way of life. Thus she remembered and clearly responded approvingly to what the priest at her Catholic college repeated every week when she was a student there: ''Get married, get pregnant. Get married, get pregnant. Then your husband will mature and be doing the most masculine of acts—supporting you. And you will be doing the most feminine of acts—nurturing. And you will mature. And you will be secure'' (Klatch 1982, p. 173). Not incidentally she also identifies with the Moral Majority. It ''is only saying what the Catholic Church has been saying all along. . . . Now the Moral Majority is doing a lot of guiding where it is needed'' (Klatch 1982, p. 168).

Donnelly's political activism was aroused by the *Roe* decision and soon extended to the ERA, which she opposed on its own terms, but also because of its link to abortion, alleging that under the Massachusetts ERA public funds were mandated to pay for abortions for poor people. Her political concerns included the church itself, where she deplored the fact, as she saw it (Klatch 1982, p. 168), that the ''nuns today'' want ''to do mass and everything.''

Let us now look at the case of someone who, brought up as a Catholic, left the faith completely and went on to become a prominent feminist. Kate Millett grew up in St. Paul, Minnesota. The city was overwhelmingly immigrant Irish; most "of these staunchly Catholic children, Kate among them, attended parochial school with pride" (Cohen 1988, p. 73). The homes of her relatives on her father's side had a decidedly intellectual, cultural bent, but books "were important at home as well." As she matured, Kate fully shared this cosmopolitan world. "Soon, like the sophisticated Millett side of the family, Kate was making fun of more recent doctrines such as the Immaculate Conception and the Assumption" (Cohen 1988, p. 75). Despite the displeasure of her mother, and the pervasive authority of the Church in her early life, she persisted in questioning its authority. And quite contrary to Mary Donnelly, who clearly was always comfortable with the traditional separation of sex roles within the Church, Millett, even as a school girl, deplored what she perceived to be and what would later be labeled as sexism in its structure. Her intellectual questioning continued as Millett attended the University of Minnesota and as, following graduation, she earned first class honors at Oxford; the break with the Church was, at the same time, encouraged by her incipient lesbian relationships.

Three brief comments conclude this section. First, although admittedly no one is born a fundamentalist Protestant or a devout Catholic, people may be born with varying predispositions that lend themselves to particular religious creeds and practices. Thus, women with a strong desire for children who accordingly raise large families may find the pro-motherhood, pro-family message of Catholicism and fundamentalist Protestantism quite congenial; indeed, the religious doctrines may justify, fortify, and magnify (hypertrophy) these preexisting tendencies. Dynamic self-selection is involved, for as Luker (1984) indicated, nearly 20 percent of her pro-life activists converted to Catholicism, while fully that proportion left the Church of their childhood, becoming pro-choice activists and clearly lacking the pro-family orientation of their opponents. It is of interest that the Minnesota Study of Twins (Bouchard et al., 1990) indicates a genetic component in the formation of key religious attitudes.

Second, religion is not a monolithic entity. It is sufficiently vast and varied to accommodate the more cosmopolitan as well as the more local. Accordingly, Durkheim (1965) can stress its local and collective nature, James (1958) and Allport (1971), its individualistic and intellectual character. Protestantism ranges from the cosmopolitan Unitarian-Universalists to the fundamentalist Protestants, whose literal reading of Scripture fits a localist frame of mind. The Catholic Church includes sophisticated theologians who would be more comfortable talking with their Protestant counterparts than a grass-roots Catholic parish women's group. Finally, much of the supposed contradiction between the universal message of peace and brotherhood of the major religions and the parochial practice and xenophobic conflict of most of their sects may result from the fact that the universal message is highlighted by a small minority of cosmopolitan writers. It is precisely the mark of a cosmopolitan to transcend a narrow sectarianism, although a subjectively trapped cosmopolitan may be prone to a secular, though quasi-religious, "ism."

Thirdly, sex differences appear to correlate with religious affiliation. ''One of the best-known findings in the psychology of religion is the greater religiosity of women, in the Christian world at least'' (Argyle 1985, p. 10). Argyle notes the absence of any conclusive explanation for this result, but he does allude in this connection to the closer social ties formed by women, including within families, and the biological predispositions to care for children and their families. It is noteworthy that the most active anti-ERA and pro-life women are both ardently pro-family and religious.

We can summarize in an admittedly speculative manner the possible relationship between local and cosmopolitan orientations, sex roles, and religious identification in Table 7.4.

A word is necessary to explain the use (and our qualification) of Allport's extrinsic/intrinsic distinction (1968, ch. 8), which is used in Table 7.4 to highlight their relevance to religious factors. The extrinsic religious sentiment is, for Allport, something ''used,'' not ''lived.'' It involves religion as a ''tribal'' habit, a way to ''bolster one's self-confidence,'' to ''win friends, power, or influence'' or to ''most importantly provide a super-sanction for one's own formula for living.'' Allport (1968, p. 150) clearly disapproves of this sentiment: ''In terms of developmental psychology, the formation is immature. Like Piaget's children, the possessor holds an egocentric view of the universe. Habit, custom, family tradition are not critically reevaluated and recentered into an adult outlook.''

The ''intrinsic'' religious sentiment, by contrast, involves a ''comprehensive commitment'' to ''an ideal situation of one's life, but always under a unifying conception of the nature of all existence.'' Note that this ''mature'' orientation indeed presupposes entry into Piaget's final intellectual stage of formal operations;

Table 7.4
Hypothesized Relationship between Local-Cosmopolitan Orientations, Sex Roles, and Religious Identification

Local-cosmopolitan orientation	Religious Identification	Sex Roles
Cosmopolitan (international national, sectional)	Intrinsic	Egalitarian, similar, modern (fully competitive)
Community local	Intrinsic-extrinsic	Married and family oriented females more religious (Sexually competitive)
Neighborhood local	Extrinsic	Unequal, segregated, traditional, (less competitive)

it could, in any event, hardly be couched in more explicitly cosmopolitan terms. Allport (1968, p. 151) further observes that "this conception of intrinsic religion has nothing to do with formal religious structure."

Allport (1968, p. 149) finds the extrinsic sentiment to "correlate positively with racial and ethnic bigotry. This finding would also be consistent with the sociobiological view that religious sectarianism magnifies in/out group divisions (cf. Wilson 1978). But such a position should especially apply to those capable of only limited space-time horizons.

Interestingly, in another study, Allport (1968, p. 253) reports those who are "indiscriminately pro-religious" as being the most prejudiced at all. There are subjects who have expressed agreement with *both* extrinsic and intrinsic test items, which is why I refer to this category as "extrinsic-intrinsic." This finding seems puzzling to Allport (1968, p. 254), for whom this "indiscriminate" response indicates "religious muddleheadedness."

This category, however, fits nicely with the community local orientation that also may correspond with the sexually competitive woman, etc. The more local person will fit the narrow, unreflective extrinsic religious sentiment, but the community locals, while neither intellectual nor cosmopolitan, will possess the cognitive skills to take an active interest in religious doctrine, much as Mary Donnelly did, even if it is, in Allport's terms, ultimately self-serving, that is, in bolstering her lifestyle. For such an individual, desiring children, bonding with like-minded people under religious auspices, and actively promoting congenial religious doctrines would be all of a piece.

In any case, I have plugged into this chart the same distinctions discussed earlier in order to highlight their possible relevance to religious factors. It is again appropriate to mention the temperance movement, which, in dividing the sexes, aligned the churches actively behind the American woman's attempt to subdue the saloon. The Woman's Christian Temperance Union, as the name suggests, united women and religion in defense of the family.

Territory. Two of the studies already cited find no significant differences between activists on opposing sides of ERA or abortion in place of residence. Luker's California sample (1984, p. 197) finds that the "average" pro-choice and pro-life activist "grew up in a large metropolitan area." Arrington and Kyle (1978, p. 676) state in North Carolina there "were no significant differences between the groups with respect to the region in which they were reared or how far they had moved geographically in their lifetimes." They note, however, that "they were quite distinct in terms of their attachment to the local environment."

This latter point bears emphasis. Two individuals may grow up or reside in a large metropolitan area or in an isolated rural area and yet relate to those "same" environments differently.

The dynamics of self-selection suggest some differential mobility between locals and cosmopolitans, and other data indicate this to be the case here. Most of the prominent national feminists drew away from the smaller cities of their birth, for example, Betty Friedan from Peoria, Illinois; Kate Millett from St. Paul, Minnesota;

and Gloria Steinem from East Toledo, Ohio. Conversely, Brady and Tedin (1976, p. 572) stress that nearly two-thirds of their female sample of anti-ERA activists in Texas "grew up in rural areas or in small towns" and most still reside there. Indicating that such women "are more conservative than those who grew up in the suburbs or in cities," they "suspect that provincialism is at least partially responsible for anti-ERA and associated political attitudes." They argue that "political activity could express resentment against the increasing dominance of the large city, particularly in setting standards or life styles, or it could be a nostalgia for the small town way of life." Weiss (1988, pp. 210–11) notes the clustering of core members of NOW in "affluent, cosmopolitan neighborhoods," with much greater percentages than for "rural and small-town Americans." Mansbridge (1986, p. 273) and Burris (1983, p. 309) report in their national surveys on attitudes toward ERA that significantly lower levels of support for ERA were found in rural areas. And, as Burris notes, the less urbanized states were the most overrepresented in the amendment process.

Politics. For fully competitive cosmopolitan women, political activity is a logical extension of their desire for sexual equality. Their cognitive skills encourage thinking in political terms. Just as they initiated abortion reform in California (Luker 1984) and a number of other urban states before *Roe*, they initiated the ERA. But the limitation of their cosmopolitan orientation, as already noted, is their urban, national focus; that is, their reliance on Washington- and New York-based networks that inhibit local, grass-roots organizations.

The anti-abortion activists whose mobilization was also to hurt ERA had not been, as a whole, politically active at all before *Roe* (Luker 1984, ch. 6). Two-thirds of these newer activists, according to Luker (1984, p. 147) were "self-recruits, actively seeking out "an organized political group that shared their values." Many of the others were recruited by friends who were already involved. "The friends were almost always close friends of long standing," Luker (1984, p. 149) reports, "co-workers or old classmates, neighbors, people who had children in the same school. Interestingly, many of the people who were recruited in this way reported that their friends never had a formal discussion with them about abortion but simply assumed that their anti-abortion values were shared." Luker (1984, p. 150) quotes one pro-life respondent as listing the 1973 Supreme Court decision as an impetus for her involvement along with a "friendship with a local woman I knew through school, her children were in the same school and so forth . . . and she kind of pushed me into it."

Note the implicit self-selection process at work here. Most of these women, partly because of their already large families, had been apolitical. But many of them had gravitated toward the same neighborhoods, with their children attending the same schools, previously in many cases a Catholic parochial school. The *Roe* case could then inflame such a neighborhood, where the similarity in life-style and values would be reinforcing. And a strong grass-roots organization would be the outcome of such political activation on a neighborhood level.

These activists are not neighborhood locals à la the "urban villagers." If not a college graduate, these women likely attended college and married reasonably well. Although they are narrowly focused in their pro-life cause—for many it is a "single issue"—they possess the requisite cognitive skills to mobilize effectively on a local level.

This assessment applies especially to the organized anti-ERA women. Brady and Tedin (1976, p. 572) and Arrington and Kyle (1978, pp. 670–72) find that the anti-ERA activists are *not* significantly alienated or lacking in self-esteem. In particular, they do *not* fit the feminist portrayal of them as being fearful of competition with men and therefore seeking security in their marriage and family—but nonetheless anxious concerning that security.

The feminists may be applying their fully competitive model where it does not apply. These anti-ERA women may have been sexually competitive and, in their own eyes, the winners in their competition. They gained the marriages and the families that they sought and they resent cosmopolitan attitudes that demean these successful self-selecting accomplishments. Their political mobilization, especially following the impetus of the 1973 *Roe* decision, constituted a backlash completely reversing the tide of cosmopolitan support that had been carrying ERA.

One footnote needs to be added here. Several national polls showed clear majority support for the Equal Rights Amendment. How, then, did it lose? Poll results can be misleading. As Mansbridge observes, much of the support for the amendment came from those who were championing the idea of "rights" but who were not actually committed to changing women's roles. She shows that many respondents who had clearly traditional views on sex roles nonetheless said that they favored ERA. Furthermore, nonwhite respondents were significantly more likely to say that they favored "equal rights" for women. Yet, again, such support may have been soft, vanishing when put to a test. Indeed, such may have been true for men, even among highly cosmopolitan men. Thus, Evans (1979) traces some of the roots of the contemporary women's movement in the United States to the civil rights struggle and the New Left politics of the 1960s, when the male-dominated groups involved professed (sexual) equality but did not practice it. More recently, Hochschild and Muchung (1989) observe that the most educated men in working couples also are more likely to profess equality in doing domestic chores, but are less likely to deliver than their less well-educated counterparts.

CONCLUSION

Comparing and Contrasting Perspectives

As in a sociobiological position, there is here an insistence on the significant presence of a biological dimension in understanding sex roles. But on two key points there tends to be a divergence. First, although it is well to recognize that sociobiological thinking is by no means monolithic, its tendency is to typologize differences. It is precisely this typology—women as more passive and dependent, men the reverse—that Hrdy, writing as a sociobiologist and a feminist, criticized, although her own construct of the sexually competitive female may be too typologically drawn.

Second, sociobiological thought pays insufficient attention to the intermediation of cognitive factors; these in turn are also to be placed within both an interactional and populational framework. Interests—genetic and/or social—are interpreted differently by each individual, depending in part on his or her cognitive capacity, just as the same individual may change in these respects as he or she matures from birth through adulthood.

These last two points apply equally to an environmentalist feminist position, which also tends to typologise women as potentially all fully competitive. Although this viewpoint is anything but monolithic, it also tends to downplay the importance of individual differences in cognition, especially as viewed interactionally. Finally, the most significant point of divergence concerns the strong environmentalism of the feminist position that tends to deny any biological basis, for example, underpinning the woman's role as mother following the birth of her children. (Feminists might cite, e.g., changes in fertility rates as evidence of environmental influence, which is true but does not negate a continuing biological dimension, any more than the noticeable increase in physical stature among younger Japanese following World War II does).

Historical Implications

A number of the studies cited on abortion and ERA allude to works dealing with other historical times and political issues for purposes of comparison. Let us look briefly at a few of these.

Luker (1984, p. 801) sees the pro-life movement as being a "women's version of what was true of peasants in the Vendée," whose strong ties to traditional roles led to their resistance to the French Revolution. She refers in particular to the work of Charles Tilly (1964, p. 59), who explicitly cites Robert Merton in connection with the local/cosmopolitan distinction, which Tilly reformulates as "internally" and "externally" oriented activity. In any event, the Vendée counterrevolution is viewed as a direct clash between rural-localist-traditional and urban-cosmopolitan-revolutionary elements in French society following 1789.

Another localist backlash had just occurred during the struggle for the ratification for the U.S. Constitution (see chapter 5). Note here a parallel to the ratification fight over ERA. In both cases, cosmopolitan forces dominated at the national level and had the initiative in gaining support on the state level. In both cases, however, the more localist, rural forces mobilized and hotly contested their opponents in the ratification process. Indeed, if the original constitution had to be adopted through the state legislatures, as both the Article of Confederation and its own amendment provisions required (and as ERA had to be), its ultimate success would have been more doubtful. (Later, rural localist forces were instrumental, as Berry [1986] indicates, in gaining ratification for the Sixteenth and Eighteenth Amendments to the Constitution, on the income tax and prohibition, respectively).

Arrington and Kyle (1978, p. 678) also note that their anti-ERA women activists act like "political 'Traditionals' " even though they "have the social characteristics of the modern 'Participant,' " in terms coming from Lerner's *Passing of Traditional Society* (1958). Lipset (1963, p. 44, ft. 23) notes that Lerner's key personality requirement for modern society is "empathy," which Lipset sees as a "more or less 'cosmopolitan' attribute." Although Lerner (1958, ch. 10) acknowledges extremist counterelites in places like Iran in the 1950s, he implicitly assumes a more or less unidirectional eventual "passing of traditional society." But what the revolt of the Vendée peasants in Revolutionary France anticipated is that the rise of modern secular cosmopolitan values will meet resistance, as it did indeed under the modernization program of the Shah of Iran. It needs stressing here that just as anti-ERA women may adhere to traditional roles with the cognitive skills of a community local, the same can be true for others who are traditionally minded. The working class or peasants should not be viewed as monolithic; rather, because of inevitable genetically based individual variation, a few will tend to become cosmopolitan, perhaps revolutionary leaders, if they find themselves trapped, while most others will retain a traditional life style that their more localistic leaders may reflect.

Boles (1979, p. 185) finds that with just a few qualifications, the "community conflict model accurately described many aspects of the ERA controversy": "The conflict was initially introduced into the states by a national opposition force. Ad hoc groups, usually opposed to the amendment, were formed to contest the issue." One case study illustrating the community conflict model cited by Boles is Crain, Katz and Rosenthal, *The Politics of Community Conflict: The Fluoridation Decision* (1969). Here fluoridation is shown to be initiated and supported by "professional" cosmopolitans, inspiring backlash from less well-educated "locals." Although locals and cosmopolitans surely can and do fight amongst their own numbers, one recurring historical theme seems to be one of cosmopolitan initiative followed by localist backlash.

If the cosmopolitan "fully competitive female" model held for women's issues, then with women's suffrage in a democracy, the passage of ERA would have been ensured. How could women oppose their own self-interest? But what is their self-interest? Unhappily, feminist ideals and most women's aspirations may not be identical.

A particularly graphic example may be found during the Weimar Republic, where, as Bridenthal and Koonz (1976, p. 303) write, "German women did not merely lose the fight for equality; they joined the opposition." Far from spurning the Nazi appeal to *Kinder, Küche, Kirche* ("Children, Kitchen, Church"), more women than men supported the Nazis. Bridenthal and Koonz (1976, p. 503) observe that although the Democratic party and Socialists had long advocated women's rights, most German women "voted instead for those parties which had so long opposed women's rights." The American experience with ERA must be similarly disillusioning from a feminist perspective. Not all women may wish to be liberated.

References

Acbuthnot, J., Gordon, D. A., and Jorkovic, G. S. "Personality." In *Handbook of Juvenile Delinquency*, 139–83. Ed. H. C. Quay. New York: Wiley, 1987.

Adair, Douglass. *Fame and the Founding Fathers*. Ed. T. Colburn. New York: Norton, 1974.

Adler, K., and Bobrow, D. "Interest and Influence in Foreign Affairs." *Public Opinion Quarterly* 20 (1956): 89–103.

Allport, G. W. *Personality and Social Encounter*. Boston: Beacon, 1968.

——. *The Individual and His Religion*. New York: Macmillan, 1971.

Almond, Gabriel and Verba, Sidney. *The Civil Culture*. Boston: Little Brown, 1965.

Altbach, Edith Hoshino, ed. *From Feminism to Liberation*. Cambridge, Mass.: Schenkman, 1971.

Anastasi, Anne. *Psychological Testing*. 3d ed. New York: Macmillan, 1968.

Anderson, Martin. *Revolution*. New York: Harcourt, 1988.

Archer, John, and Lloyd, Barbara. *Sex and Gender*. Cambridge: Cambridge University Press, 1985.

Argyle, Michael. "New Directions in the Psychology of Religion." In *Advances in the Psychology of Religion*. Ed. C. B. Brown. New York: Pergamon, 1985.

——. *The Psychology of Happiness*. London: Methuen, 1987.

Aristotle. *Introduction to Aristotle*. Ed. Richard P. McKeon. New York: McGraw-Hill, 1974.

Arlacchi, Pino. *Mafia Business*. London: Verso, 1986.

Arrington, T., and Kyle, P. "Equal Rights Amendment Activists in North Carolina." *Signs* 3 (1978): 666–80.

Ashworth, P. D. *Social Interaction and Consciousness*. New York: Wiley, 1979.

Ayres, B. Drummond. "A New Breed of Diplomat." *The New York Times Magazine* (September 11, 1983): 66–75.

Bachrach, P., and Baratz, M. "Decisions and Non-Decisions." *American Political Science Review* 57 (September 1963): 632–42.

Bacon, F. *The Essays*. New York: Viking Penguin.

Baer, Denise. "Gender and Groups in Politics: The Biosocial Perspective." Paper prepared for presentation at the XIVth World Congress of the International Political Science Association, August 28–September 1, 1988, Washington, D.C.

Balch, S. " 'The Neutered Civil Servant': Eunuchs, Celibates, Abductees and the Maintenance of Organizational Loyalty." In *Biology and Bureaucracy*, 271–304. Ed. E. White and J. Losco. Lanham, Md.: University Press of America, 1986.

Baldwin, J. M. "Organic Selection." *Nature*, 15 April 1897.

Banfield, Edward. *The Heavenly City*. Boston: Little, Brown, 1968.

——— . *The Heavenly City Revisited*. Boston: Little, Brown, 1974.

Banks, Stephen P., and Kahn, Michael D. *The Sibling Bond*. New York: Basic Books, 1982.

Barash, David. *The Hare and the Tortoise*. New York: Penguin, 1986.

Barber, James David. *The Presidential Character*. Englewood Cliffs, N.J.: Prentice-Hall, 1972.

Bardwick, Judith. *The Plateauing Trap*. New York: AMACOM: American Management Association, 1986.

Barnes, J. A. "Networks and Political Process." In *Social Networks in Urban Situations*, 51–76. Ed. J. Clyde Mitchell. Manchester: Manchester University Press, 1969.

——— . *Social Networks*. New York: Addison-Wesley, 1972.

Barrett, L. *Gambling with History*. New York: Penguin, 1984.

Beard, Charles A. *An Economic Interpretation of the Constitution of the United States*. 1913. Reprint. New York: Macmillan, 1961.

Bell, Robert R. *Worlds of Friendship*. Beverly Hills, Calif.: Sage, 1981.

Benbow, C. P. "Sex Differences in Mathematical Reasoning Ability in Intellectually Talented Preadolescents: The Nature, Effects, and Possible Causes." *Behavioral and Brain Sciences* 11 (1988): 169–232.

Bennis, Warren. *Why Leaders Can't Lead*. San Francisco: Jossey-Bass, 1989.

Benson, Lee. *Turner and Beard*. Glencoe, Ill.: The Free Press, 1960.

Benton, Arthur. "Spatial Thinking in Neurological Patients: Historical Aspects." In *Spatial Abilities: Development and Physiological Foundations*. Ed. M. Potegal. New York: Academic Press, 1982.

Berlioz, Hector. *Memoirs*. New York: Knopf, 1932.

Bernard, Jessie. *Journey Toward Freedom*. New York: Dell, 1967.

Berry, M. F. *Why ERA Failed*. Bloomington: Indiana University Press, 1986.

Betz, Nancy E. and Fitzgerald, Louise F. *The Career Psychology of Women*. Orlando, Fla.: Academic Press, 1987.

Bloom, Floyd, E., Lazerson, Arlyne, and Hofstadter, Laura. *Brain, Mind, and Behavior*. New York: W. H. Freeman, 1985.

Boehm, Christopher. "Rational Preselection from Hamadryas to *Homo Sapiens*: The Place of Decisions in Adaptive Process." *American Anthropologist* 80 (1978): 265–96.

Boles, Janet K. *The Politics of the Equal Rights Amendment*. New York: Longman, 1979.

Borden, Morton, ed. *George Washington: Great Lives Observed*. Englewood Cliffs, N.J.: Prentice-Hall, 1969.

Bott, E. *Family and Social Network*. New York: The Free Press, 1971.

Bouchard, T. J., et al. "Sources of Human Psychological Differences: The Minnesota Study of Twins Reared Apart." *Science* 250 (1990): 223–28.

Bradlee, Ben Jr. *Guts and Glory*. New York: Donald F. Fine, 1988.
Brady, D., and Tedin, K. ''Ladies in Pink: Religion and Political Ideology in the Anti-ERA Movement.'' *Social Sciences Quarterly* 56 (1976): 564–75.
Brain, Robert. *Friends and Lovers*. New York: Basic Books, 1976.
Brant, Irving. *James Madison: The Virginia Revolutionist*. New York: Bobbs-Merrill, 1941.
——. *James Madison: The Nationalist: 1780–1787*. New York: Bobbs-Merrill, 1948.
Bridenthal, R., and Koonz, C. ''Beyond *Kinder, Küche, Kirche*: Weimar Women in Politics and Work.'' In *Liberating Women's History*. Ed. B. Carroll. Urbana: University of Illinois Press, 1976.
Broder, D. S. *Changing of the Guard*. New York: Simon & Schuster, 1980.
Brown, P. J. and Arnold, T. E. ''Counterterrorism as Enterprise: The Iran-Contra Investigations Spotlight the New Private Sector.'' In *Beyond the Iran-Contra Crisis*, 191–209. Eds. P. C. Livingstone and T. E. Arnold. Lexington, Mass.: Lexington, 1988.
Brown, Robert C. *Charles Beard and the Constitution*. Princeton, N.J.: Princeton University Press, 1956.
Bruner, Jerome. *In Search of Mind*. New York: Harper Colophon, 1984.
Brunhouse, Robert L. *The Counter-Revolution in Pennsylvania: 1776–1790*. Harrisburg: Pennsylvania Historical Commission, 1942.
Brzezinksi, Zbigniew. *Power and Principle*. New York: Farrar, Straus and Giroux, 1983.
Burns, J. M. *Congress on Trial*. New York: Harper, 1949.
Burris, Val. ''Who Opposed the ERA? An Analysis of the Social Bases of Anti-Feminism.'' *Social Science Quarterly* 64 (1983): 305–10.
Burt, R. S. and Minor, M. S. *Applied Network Analysis*. Beverly Hills, Calif.: Sage, 1983.
Campbell, Angus, et al. *The American Voter*. New York: Wiley, 1964.
Carrol, Susan J. ''Gender Politics and the Socializing Impact of the Women's Movement.'' In *Political Learning in Adulthood* 306–39. Ed. R. Sigel. Chicago: University of Chicago, 1989.
Casey, William. *The Secret War Against Hitler*. Washington, D.C.: Regnery Gateway, 1988.
Caton, Hiram. ''The Social Technology Model of Political Organization.'' Paper delivered at the American Political Science Association annual meeting, Chicago, September 1982.
——. ''Feminism and the Family.'' Cleveland, Queensland, Australia: The Council for a Free Australia, 1985.
Cattell, R. B. *The Scientific Analysis of Personality*. Chicago: Aldine, 1965.
——. *The Inheritance of Personality and Ability*. New York: Academic Press, 1982.
Chafe, William H. *The American Women*. New York: Oxford University Press, 1972.
Chafetz, Janet Saltzman and Dworkin, Anthony Gary. *Female Revolt*. Towota, N.J.: Rowman & Allanhold, 1986.
Chaplin, Charles. *My Autobiography*. New York: Simon & Schuster, 1974.
Charlesworth, William R. ''An Ethological Approach to Cognitive Development.'' In *Recent Advances in Cognitive Development Research*, 237–58. Ed. Charles J. Brainerd. Berlin: Springer-Verlag, 1983.
Chisholm, Shirley. *Unbought and Unsold*. New York: Houghton Mifflin, 1971.
Chodorow, Nancy. ''Feminism and Difference: Gender, Relation, and Difference in Psychoanalytic Perspective.'' In *The Psychology of Women*, 249–64. Ed. Mary Ruth Walsh. New Haven, Conn.: Yale University Press, 1987.
Churchill, Winston. *The Gathering Storm*. Boston: Houghton Mifflin, 1948.
——. *Their Finest Hour*. Boston: Houghton Mifflin, 1949.

Cohen, Gillian. *The Psychology of Cognition*. 2d ed. London: Academic Press, 1983.
Cohen, Marcia. *The Sisterhood*. New York: Simon & Schuster, 1988.
Colby, William. *Honorable Men*. New York: Simon & Schuster, 1978.
Collins, Nancy W., Gilbert, Susan K., and Nycum, Susan H. *Women Leading*. Lexington, Mass.: The Stephen Greene Press, 1988.
Conover, P., and Gray, V. *Feminism and the New Right*. New York: Praeger, 1983.
Conquest, R. *The Great Terror*. New York: Oxford, 1990.
Conway, J., Bourgue, S. C., and Scott, J. W. "Introduction: The Concept of Gender." *Learning About Women: Gender, Politics, and Power, Daedalus* 116 (Fall 1987): v–xxix.
Cooper, H.H.A., and Redlinger, L. J. *Making Spies*. Boulder, Colo.: Paladin Press, 1986.
Corning, Peter A. *The Synergism Hypothesis: A Theory of Progressive Evolution*. New York: McGraw-Hill, 1983.
Cosell, H. *Woman on a Seesaw*. New York: G. P. Putnam's, 1985.
Cottle, Thomas S., and Klineberg, Stephen L. *The Present of Things Future*. New York: The Free Press, 1974.
Cox, Catherine M. *Genetic Studies of Genius*. Vol. II, *The Early Mental Traits of Three Hundred Geniuses*. Stanford, Calif.: Stanford University Press, 1926.
Craig, Albert. "Introduction." In *Perspectives on Personality in Japanese History*. Ed. A. M. Craig and D. H. Shively. Berkeley: University of California, 1970.
Crain, R., Katz, E., and Rosenthal, D. *The Politics of Community Conflict: The Fluoridation Decision*. New York: Irvington, 1969.
Crenson, M. "Social Networks and Political Processes in Urban Neighborhoods." *American Journal of Political Science* 22 (1978): 578–95.
Csikszentmihalyi, M. *Flow: The Psychology of Optimal Experience*. New York: Harper, 1990.
Darlington, C. D. *Genetics and Man*. New York: Schocken, 1969.
——. *The Evolution of Man and Society*. New York: Simon & Schuster, 1971.
Davies, J. C. "Ions of Emotion and Political Behavior." "In *Biology and Politics*. Ed. A. Somit. Paris: Mouton, 1976.
Davis, Joseph L. *Sectionalism in American Politics: 1774–1787*. Madison: University of Wisconsin Press, 1977.
Dawkins, R. *The Selfish Gene*. New York: Oxford University Press, 1976.
de Beauvoir, Simone. *The Second Sex*. New York: Knopf, 1953.
——. *The Prime of Life*. New York: Lancer Books, 1962.
Degler, C. *At Odds*. New York: Oxford, 1980.
——. *In Search of Human Nature, the Decline and Revival of Darwinism in American Social Thought*. New York: Oxford University Press, 1991.
de Waal, F.B.M. "Sex Differences in the Formation of Coalitions among Chimpanzees." *Ethology and Sociobiology* 5 (1984): 239–55.
——. *Peacemaking among Primates*. Cambridge, Mass.: Harvard University Press, 1989.
Diamond, Martin, et al. *The Democratic Republic*. Chicago: Rand McNally, 1972.
Dick, Steven W. *Person Relationships and Personal Constructs: A Study of Friendship Formation*. New York: Wiley, 1973.
The Dictionary of American Biography. Vol. VI. New York: Charles Scribner's Sons, 1959.
Dobzhansky, Theodosius. *Mankind Evolving*. New York: Bantam, 1964.
——. *Genetics of the Evolutionary Process*. New York: Columbia University Press, 1970.
——. *Genetic Diversity and Human Equality*. New York: Basic Books, 1973.

Dobzhansky, Theodosius, and Boesiger, Ernest. *Human Culture: A Moment in Evolution*. New York: Columbia University Press, 1983.

Douglass, Frederick. *Narrative*. Garden City, N.Y.: Dolphin Books, 1963.

Downs, Roger M. "Maps and Mappings as Metaphors for Spatial Representation." In *Spatial Representation and Behavior Across the Life Span*, 143–66. Ed. L. Liker, A. H. Patterson and N. Newcombe. New York: Academic Press, 1981.

Drucker, Peter F. *Management*. New York: Harper & Row, 1974.

Dulles, Allen. *The Craft of Intelligence*. New York: Harper & Row, 1963.

Dunn, L. C., and Dobzhansky, T. *Heredity, Race and Society*. New York: Mentor, 1952.

Durkheim, Émile. *The Elementary Forms of Religious Life*. New York: The Free Press, 1965.

Dye, T. "The Local-Cosmopolitan Dimension and the Study of Urban Politics." *Social Forces* 41 (1966): 239–46.

Eccles, J. *The Understanding of the Brain*. 2d ed. New York: McGraw-Hill, 1977.

Edelman, Gerald. *Neuronal Darwinism*. New York: Basic Books, 1987.

Eichenbaum, L., and Orbach, S. *Between Women*. New York: Viking, 1988.

Elkins, Stanley, and McKitrick, Eric. "The Founding Fathers: Young Men of the Revolution." In *The Formation of the Constitution*, 65–75. Ed. R. F. Jones. New York: Robert C. Krieger, 1978.

Ellis, Rafaela. *The Central Intelligence Agency*. New York: Chelsea House, 1988.

Emerson, Ralph Waldo. *Essays*. New York: Vintage, 1961.

Emerson, Steven. *Secret Warriors*. New York: G. P. Putnam's, 1988.

Emmert, M. "Organizational Cohesion and Commitment: An Evolutionary Perspective. In *Biology and Bureaucracy*, 233–70. Ed. E. White and J. Losco. Lanham, Md.: University Press of America, 1986.

Etkin, William. "The Expendable Male." *Perspectives in Biology and Medicine* 22 (1979): 559–64.

——. "A Biological Critique of Sociobiology." In *Sociobiology and Human Politics*. Ed. E. White. Lexington, Mass.: Lexington Books, 1981.

——. "Evolution of the Human Mind and Emergence of Tribal Culture: A Mentalist Approach. *Perspectives in Biology and Medicine* 28, no. 4 (1985): 498–525.

Evans, Sara. *Personal Politics*. New York: Knopf, 1979.

Ferguson, Russell. *Early Western Pennsylvania Politics*. Pittsburgh, Pa.: University of Pittsburgh Press, 1938.

Figueira-McDonough, Josefina, and Sarri, Rosemary, eds. *The Trapped Woman*. Newbury Park, Calif.: Sage, 1987.

Findley, William. "An Autobiographical Letter." *Pennsylvania Magazine of History and Biography* 5 (1881): 440–50.

Flavell, J. H. *The Developmental Psychology of Jean Piaget*. Princeton, N.J.: Van Nostrand, 1963.

Flexner, James Thomas. *George Washington in the American Revolution: 1775–1783*. Boston: Little, Brown, 1967.

——. *The Young Hamilton*. Boston: Little, Brown, 1978.

——. *Washington: The Indispensable Man*. New York: Signet, 1984.

Forster, E. M. "I Believe." In *A Quarto of Modern Literature*, 625–29. Ed. L. Brawn and P. G. Perrin. New York: Charles Scribner's, 1950.

Fox, Mary Frank. "Women in the Labor Force: Position, Plight, Prospects." In *The Trapped Woman*, 197–215. Ed. J. Figueira-McDonough and R. Sarri. Newbury Park, Calif.: Sage, 1987.

Fraiberg, Selma. *Every Child's Birthright: In Defense of Mothering*. New York: Basic Books, 1977.
Fraisse, Paul. *The Psychology of Time*. Westport, Conn.: Greenwood, 1963.
Frankl, Viktor. *Man's Search for Meaning*. New York: Pocket Books, 1984.
Franklin, Benjamin. *Autobiography*. New York: Washington Square, 1966.
Freeman, Derek. *Margaret Mead and Samoa: The Making and Unmaking of an Anthropological Myth*. Cambridge, Mass.: Harvard University Press, 1983.
Freeman, Douglas S. *George Washington*. Vol. II. New York: Charles Scribner's, 1948.
Freeman, Jo. *The Politics of Women's Liberation*. New York: McKay, 1975.
Friedan, Betty. *The Feminine Mystique*. New York: Norton, 1974.
——. *The Second Stage*. New York: Summit, 1981.
——. *It Changed My Life*. New York: Norton, 1985.
Galbraith, J. K. *The New Industrial State*. New York: Signet, 1967.
Gans, Herbert J. *The Urban Villagers*. Glencoe, Ill.: The Free Press, 1965.
Gardner, Howard. *Frames of Mind*. New York: Basic Books, 1983.
Gatlin, Rochelle. *American Women Since 1945*. Jackson: University Press of Mississippi, 1987.
Genevie, L., and Margolies, E. *The Motherhood Report*. New York: Macmillan, 1987.
Gilligan, Carol. *In a Different Voice*. Cambridge, Mass.: Harvard University Press, 1982.
Ginzberg, Eli, and Yohalem, Alice M. *Educated American Women: Self-Portraits*. New York: Columbia University Press, 1966.
Goertzel, Mildred George, Goertzel, Victor, and Goertzel, Ted George. *Three Hundred Eminent Personalities*. San Francisco: Jossey-Bass, 1978.
Goldstein, Kurt. *Human Nature in the Light of Psychopathology*. New York: Schocken, 1963.
Goldstein, Kurt, and Scheerer, Martin. "Abstract and Concrete Behavior." *Psychological Monographs*, Vol. XXXIII, 1941.
Goldstone, Bazelon. *Trading Up*. New York: E. P. Dutton, 1988.
Goodall, Jane. *The Chimpanzees of Gombe*. Cambridge, Mass.: The Belknap Press, 1986.
Gorman, Bernard S., and Weissman, A. E. "Images, Values and Concepts of Time in Psychological Research." In *The Personal Experience of Time*, 217–63. Ed. Bernard S. Gorman and A. E. Weissman. New York: Plenum, 1977.
Gould, S. J. *The Mismeasure of Man*. New York: Norton, 1981.
——. *Wonderful Life*. New York: Norton, 1990.
Granit, R. *The Purposive Brain*. Cambridge, Mass.: MIT Press, 1977.
Gulick, Luther. "Democracy and Administration Face the Future." Lecture delivered at Indiana University, Bloomington, Indiana, April 1, 1976.
Gurvitch, Georges. *The Spectrum of Social Time*. Dordrecht, Holland: Reidel, 1964.
Gustavus, S., and Henley, J. "Correlates of Voluntary Childlessness in a Select Population." In *Pronatalism*. Ed. E. Peck and J. Senderowitz. New York: Crowell, 1974.
Gutman, Roy. *Banana Diplomacy*. New York: Simon & Schuster, 1988.
Haldeman, H. R. *The Ends of Power*. New York: New York Times Books, 1978.
Handler, P., ed. *Biology and the Future of Man*. New York: Oxford University Press, 1970.
Hardy, Alister. *The Living Stream: Evolution and Man*. New York: Meridian, 1965.
Harvey, O. S., Hart, O. S., and Schroeder, H. M. *Conceptual Systems and Personality Organization*. New York: Wiley, 1961.
Haskins, Charles. *The Renaissance of the 12th Century*. Cleveland: Meridian, 1961.
Heclo, Hugh. "Issue Networks and the Executive Establishment." In *The New American Political System*, 87–125. Ed. A. King. Washington, D.C.: American Enterprise Institute for Public Policy, 1978.

Hewlett, Sylvia Ann. *A Lesser Life*. New York: Morrow, 1986.
Hinde, Robert A. *Towards Understanding Relationships*. London: Academic Press, 1979.
——. "Attachment: Some Conceptual and Biological Issues." In *The Place of Attachment in Human Behavior*, 60–78. Ed. Colin Murray Parkes and Joan Stevenson. New York: Basic Books, 1982.
Hines, S. "Biopolitics and the Evolution of Inquiry in Political Science." *Politics and the Life Sciences* 1, no. 1 (July 1982): 5–37.
Hirsch, J. "Behavior Genetic Analysis and its Biosocial Consequences." *Seminar in Psychology* 2 (1972): 97.
Hitler, Adolf. *Mein Kampf*. Boston: Houghton Mifflin, 1954.
Hochschild, Arlie. "Making It: Marginality and Obstacles to Minority Consciousness." In *Women and Science* 194–99. Ed. Ruth Kundsin. New York: Morrow, 1974.
Hochschild, Arlie, and Machung, Anne. *The Second Shift*. New York: Viking Penguin, 1989.
Hoffer, Eric. *The True Believer*. New York: Perennial, 1981.
Hole, Judith, and Levine, Ellen. *The Rebirth of Feminism*. New York: Quadrangle, 1971.
Houghan, Jim. *Spooks*. New York: Morrow, 1978.
——. *Secret Agenda*, 231–50. New York: Random House, 1984.
Hrdy, Sarah Blaffer. *The Women That Never Evolved*. Cambridge, Mass.: Harvard University Press, 1981.
Humphreys, Lloyd. "General Intelligence." In *Handbook of Intelligence*, 201–24. Ed. Benjamin Wolman. New York: Wiley, 1985.
Hunt, E. Howard. *Undercover*. New York: Berkeley Publishing Corp., 1974.
Hutter, Horst. *Politics as Friendship*. Waterloo, Ont.: Wilfrid Laurier University Press, 1978.
Hynd, George W., and Willis, W. Grant. "Neurological Foundations of Intelligence." In *Handbook of Intelligence*, 119–57. Ed. Benjamin B. Wolman. New York: Wiley, 1985.
Jackson, George. *Soledad Brother*. New York: Bantam, 1970.
James, William. *Varieties of Religious Experience*. New York: Mentor, 1958.
Janis, Irving L. *Victims of Groupthink*. Boston: Houghton Mifflin, 1972.
——. "Groupthink in Washington." *New York Times*, May 28, 1962, 1973.
——. *Groupthink*. Boston: Houghton Mifflin, 1982.
——. *Crucial Decisions*. New York: The Free Press, 1989.
Jay, Anthony. *Corporation Man*. New York: Random House, 1971.
Jencks, Christopher, et al. *Inequality*. New York: Basic Books, 1972.
Jensen, Merrill. *The New Nation*. New York: Knopf, 1958.
Jerison, Harry J. *Evolution of the Brain and Intelligence*. New York: Academic Press, 1973.
——. "The Evolution of the Brain." In *The Human Brain*, 39–62. Ed. M. C. Wittrock. Englewood Cliffs, N.J.: Prentice-Hall, 1977.
Kaplan, Robert M. "The Controversy Related to the Use of Psychological Tests." In *Handbook of Intelligence*, 465–504. Ed. Benjamin B. Wolman. New York: Wiley, 1985.
Katz, Jack. *Seductions of Crime: The Moral and Sensual Attractions in Doing Evil*. New York: Basic Books, 1988.
Kauffman, Herbert. *The Forest Ranger*. Baltimore, Md.: Johns Hopkins University Press, 1960.
Kellerman, Barbara. *All the President's Kin*. New York: The Free Press, 1981.
Keniston, Kenneth. *The Young Radicals*. New York: Harcourt, 1968.

Kissinger, Henry. *Years of Upheaval*. Boston: Little, Brown, 1982.
Kitcher, Philip. *Vaulting Ambition*. Cambridge, Mass.: MIT Press, 1985.
Kitchin, William. "The Dialectic of Dereliction: A Neurolinguistic Analysis of the Language of the Watergate Group." In *Biology and Bureaucracy*. Ed. E. White and J. Losco. Lanham, Md.: University Press of America, 1986.
Klatch, P. *Women of the New Right*. Philadelphia: Temple University Press, 1982.
Klein, Ethel. *Gender Politics*. Cambridge, Mass.: Harvard University Press, 1984.
Klein, Herb. *Making It Perfectly Clear*. New York: Doubleday, 1980.
Klugman, Jeffrey. *The New Soviet Elite*. New York: Praeger, 1990.
Knightley, Phillip. *The Second Oldest Profession*. New York: Norton, 1986.
Koch, Helen. *Twins and Twin Relations*. Chicago: University of Chicago Press, 1966.
Kohen-Raz, Reuven. *The Psychobiological Aspects of Cognitive Growth*. New York: Academic Press, 1977.
Kohlberg, L. "Stage and Sequence: The Cognitive-Descriptive Approach to Socialization." In *Handbook of Socialization Theory and Research*, 347–480. Ed. D. A. Goslin. Chicago: Rand McNally, 1969.
Kohlberg, L., and Gilligan, C. "The Adolescent as a Philosopher: The Discovery of the Self in a Postconventional World." *Daedalus* 100 (1971): 1051–86.
Komarovsky, M. *Blue-Collar Marriage*. New York: Vintage, 1967.
Kort, Fred. "An Evolutionary-Neurobiological Explanation of Political Behavior and the Lumsden-Wilson 'Thousand-Year Rule'." *Journal of Social and Biological Structures* 6 (1983): 218–30.
Lane, Robert. *Political Ideology*. Glencoe, Ill.: The Free Press, 1962.
Laqueur, Walter. *A World of Secrets*. New York: Basic Books, 1985.
Ledeen, Michael. *Perilous Statecraft*. New York: Charles Scribner's, 1988.
Lefebvre, Georges. *The Framing of the French Revolution*. New York: Vintage, 1959.
Lerner, Daniel. *The Passing of Traditional Society*. New York: The Free Press, 1958.
Levy, Jerre. "Origins of Lateral Asymmetry." In *Lateralization in the Nervous System*. Ed. Stevan Harnad. New York: Academic Press, 1977.
Lewis, Michael. *Liar's Poker*. New York: Norton, 1989.
Libby, Orin G. *The Geographical Distribution of the Vote of the Thirteen States on the Federal Constitution*. Madison: University of Wisconsin, 1894.
Liddy, Gordon. *Will*. New York: Dell, 1980.
Lipset, S. M. *Political Man*. New York: Anchor, 1963.
Livingston, J. Sterling. "Pygmalian in Management." In *Harvard Business Review: On Human Relations*, 185–94. New York: Harper and Row, 1974.
Loden, Marilyn. *Feminine Leadership*. New York: Times Books, 1985.
Losco, J. "Evolution, Consciousness, and Political Thought." Paper presented at the 1982 annual meeting of the American Political Science Association, Denver, Colorado, 1982.
Lukas, J. Anthony. *Nightmare: The Underside of the Nixon Years*. New York: Viking, 1976.
Luker, Kristin. *Abortion and the Politics of Motherhood*. Berkeley: University of California Press, 1984.
Lumsden, Charles, and E. W. Wilson. *Genes, Mind and Culture*. Cambridge, Mass.: Harvard University Press, 1981.
——. *Promethean Fire*. Cambridge, Mass.: Harvard University Press, 1983.
Lykken, David T. "Research with Twins: The Concept of Emergenesis." Presidential address delivered at the annual meeting of the Society for Psychophysiological Research, Washington, D.C., October 31, 1981.

Lyndon, Alice Atkinson. ''Juxtapositions.'' In *Working It Out*, 25–37. Ed. Sara Ruddick and Pamela Daniels. New York: Pantheon, 1977.

Maas, Peter. *Manhunt*. New York: Random House, 1986.

Maccoby, Eleanor. *The Development of Sex Differences*. Stanford, Calif.: Stanford University Press, 1965.

Maccoby, Eleanor Emmons, and Jacklin, Carol Nagy. Stanford, Calif.: Stanford University Press, 1974.

Maccoby, Michael. *The Gamesman*. New York: Simon & Schuster, 1976.

MacDonald, Kevin B. *Social and Personality Development*. New York: Plenum, 1988a.

——. ''Sociobiology and the Cognitive-Developmental Tradition in Moral Development Research.'' In *Sociobiological Perspectives in Human Development*, 140–63. Ed. K. MacDonald. New York: Springer Verlag, 1988b.

Madison, James. *Notes of Debates in the Federal Constitution*. New York: Norton, 1987.

Madsen, D. ''A Biochemical Property Relating to Power-Seeking in Humans.'' *American Political Science Review* 79 (1985): 448–57.

Main, Jackson Turner. *The Antifederalists: Critics of the Constitution*. Chapel Hill: University of North Carolina Press, 1961.

——. *The Sovereign States: 1775–1783*. New York: Franklin Watts, 1973.

Malcolm X. *Autobiography*. New York: Grove, 1966.

Malmberg, Torsten. *Human Territoriality*. The Hague: Martin, 1980.

Mannheim, Karl. *Man and Society in an Age of Reconstruction*. London: Routledge & Kegan, 1940.

Mansbridge, Jane J. *Why We Lost the ERA*. Chicago: University of Chicago Press, 1986.

Marchetti, Victor, and Marks, John. *The CIA and the Cult of Intelligence*. New York: Knopf, 1974.

Marshall, Jonathan, Scott, Peter D., and Hunter, Jane. *The Iran-Contra Connection*. Boston: South End, 1987.

Masters, Roger. ''Why Bureaucracy?'' In *Biology and Bureaucracy*. Ed. E. White and J. Losco. Lanham, Md.: University Press of America, 1986.

——. *The Nature of Politics*. New Haven, Conn.: Yale University Press, 1989.

Matthews, Donald G., and DeHart, Jane S. *Sex, Gender, and the Politics of ERA*. New York: Oxford University Press, 1990.

Maurer, David W. *The American Confidence Man*. Springfield, Ill.: Charles C. Thomas, 1974.

Mayer, J., and McManus, D. *Landslide*. Boston: Houghton Mifflin, 1988.

Mayr, Ernst. *Animal Species and Evolution*. Cambridge, Mass.: Harvard University Press, 1963.

——. ''Biological Man and the Year 2000.'' In *Toward the Year 2000: Work in Progress. Daedalus* (Summer 1967): 832–36.

——. ''The Nature of the Darwinian Revolution.'' *Science* 176 (1972): 981–89.

——. *The Growth of Biological Thought: Diversity, Evolution and Inheritance*. Cambridge, Mass.: The Belknap Press, 1982.

——. *Toward a New Philosophy of Biology*. Cambridge, Mass.: The Belknap Press, 1988.

McDonald, Forrest. *We The People*. Chicago: University of Chicago Press, 1963.

McGee, Mark G. ''Spatial Abilities: The Influence of Genetic Factors.'' In *Spatial Abilities: Development and Psychological Foundations*, 199–222. Ed. M. Potegal. New York: Academic Press, 1982.

McMaster, John Bach, and Stone, Frederick D. *Pennsylvania and the Federal Constitution 1787–1788*. Vol. 2. New York: DaCapo Press, 1970.

Mead, Margaret. *Continuities in Cultural Evolution*. New Haven, Conn.: Yale University Press, 1964.

Melore, M., and Lettick, A. "Sex Education at Benhaven." In *Autism in Adolescents and Adults*. Ed. E. Schopler and G. Mesibov. New York: Plenum, 1981.

Meloy, J. Reid. *The Psychopathic Mind*. Northvale, N.J.: Jason Aronson, 1988.

Menges, Constantine. *Inside the National Security Council*. New York: Simon & Schuster, 1988.

Merkl, Peter H. *Political Violence under the Swastika*. Princeton, N.J.: Princeton University Press, 1975.

Merton, Robert K. *Social Theory and Social Structure*. Glencoe, Ill.: The Free Press, 1957.

Meyers, Marin. *The Mind of the Founder: Sources of the Political Thought of James Madison*. Hanover, N.H.: University Press of New England, 1981.

Miller, John C. *Alexander Hamilton: Portrait in Paradox*. New York: Harper & Brothers, 1959.

Minnhaar, Gretchen. "Architect." In *Women and Success*, 25–31. Ed. Ruth Kundsin. New York: Morrow, 1974.

Mitchell, Broadus. *Alexander Hamilton: Youth to Maturity 1755–1788*. New York: Macmillan, 1957.

Morgan, C. H. *Habit and Instinct*. London: Arnold, 1896.

Morris, Richard B. *The Basic Ideas of Alexander Hamilton*. New York: Pocket Books, 1957.

——— . "Benjamin Franklin's Grand Design." In *Times of Trial*. Ed. Allan Nevins. New York: Knopf, 1958.

Mullen, Robert K. "Low-Intensity Terrorism Against Technological Infrastructures." In *Beyond the Iran-Contra Crisis*, 141–54. Ed. N. C. Livingstone and T. E. Arnold. Lexington, Mass.: Lexington Books, 1988.

Nakane, Chie. *Japanese Society*. London: Weidenfeld & Nicolson, 1970.

——— . *Human Relations in Japan*. Tokyo: Ministry of Foreign Relations, 1972.

Neff, Walter S. *Work and Human Behavior*. 3rd ed. New York: Aldine, 1985.

Neubauer, P. B., and Neubauer, Alexander. *Nature's Thumbprint*. Reading, Mass.: Addison-Wesley, 1990.

Nixon, Richard. *Six Crises*. New York: Doubleday, 1962.

——— . *Memoirs*. New York: Grosset and Dunlap, 1978.

Nobel, Mary. "Social Network: Its Use as a Conceptual Framework in Family Analysis." In *Network Analysis: Studies in Human Interaction*, 3–13. Ed. Jeremy Boisserain and J. Clyde Mitchell. The Hague: Moutin, 1973.

North, Oliver. *Taking the Stand*. New York: Pocket Books, 1987.

Ogburn, W. F. *Social Change*. New York: Viking, 1922.

Olson, David, and Bialystok, Ellen. *Spatial Cognition*. Hillsdale, N.J.: Lawrence Erlbaum, 1983.

Orme, J. E. *Trial, Experience and Behavior*. London: Life Books, 1969.

Ornstein, R. C. *On the Experience of Time*. Middlesex, Eng.: Penguin, 1969.

Perry, Lewis. *Intellectual Life in America*. New York: Franklin Watts, 1984.

Peter, L., and Hull, R. *The Peter Principle*. New York: Morrow, 1969.

Peterson, S. A. "The Human Brain and Hypostatizing." Paper presented at the meeting of the International Society for Political Psychology, Washington, D.C., 1982.

Piaget, Jean. "The Development of Time Concepts in the Child." In *Psychology of Childhood*, 34–44. Ed. P. Hoch and J. Rubin. New York: Grune & Stratton, 1955.

——. "Time Perception in Children." In *The Voices of Time*, 31–55. Ed. J. T. Fraser. New York: Braziller, 1966.

Piaget, Jean, and Inhelder, Barbel. *The Child's Conception of Space*. London: Routledge & Kegan Paul, 1956.

Plato. *Dialogues*. 2 vols. Trans. B. Jowett. New York: Random House, 1957.

Poole, Keith, T. and Zeigler, L. Harmon. *Women, Public Opinion and Politics*. New York: Longman, 1985.

Popenoe, P. "Motivation of Childless Marriages." In *Pronatalism*. Ed. E. Peck and J. Senderowitz. New York: Crowell, 1974.

Popper, Karl. *Objective Knowledge*. Clarendon: Oxford University Press, 1979.

Popper, Karl and Eccles, J. *The Self and Its Brain*. Berlin: Springer-Verlag, 1977.

Powers, Thomas. *The Man Who Kept the Secrets*. New York: Knopf, 1979.

Prados, John. *Presidents' Secret Wars*. New York: Morrow, 1986.

Pribram, K. "Self-consciousness and Intentionality." In *Consciousness and Self-Regulation*. Ed. G. E. Schwartz and P. Shapiro. New York: Plenum, 1976.

——. "Holographic Memory." *Psychology Today* 12 (1979): 75–86.

——. "The Brain as the Locus of Cognitive Controls in Action." In *Cognition in Human Motivation*. Ed. G. d'Ydewalle and W. Lens. Louvain, Belgium: Leuven University Press, 1981.

Prouty, L. Fletcher. *The Secret Team*. Englewood, N.J.: Prentice-Hall, 1973.

Quarles, Benjamin. *Frederick Douglass*. New York: Atheneum, 1968.

Quay, Herbert C. "Intelligence." In *Handbook of Juvenile Delinquency*. Ed. H. C. Quay. New York: Wiley, 1987a.

——. "Patterns of Delinquent Behavior." In *Handbook of Juvenile Delinquency*. Ed. H. C. Quay. New York: Wiley, 1987b.

Ra'anan, Uri, and Lukes, Igor, eds. *Inside the Apparat*. Lexington, Mass.: Lexington Books, 1990.

Raine, Adrian. "Antisocial Behavior and Social Psychophysiology." In *Social Psychophysiology and Emotions*. Ed. H. L. Wagner. New York: Wiley, 1988.

Ranelagh, John. *The Agency*. New York: Simon & Schuster, 1986.

Rather, Dan, and Gates, Paul. *The Palace Guard*. New York: Harper & Row, 1974.

Ravich, Robert A. "A Sort of 'Natural Commune.' " *The New York Times Magazine*, November 5, 1972, 75–87.

Regan, Donald. *For the Record*. New York: Harcourt, Brace, Jovanovich, 1988.

Reich, Robert. *The Work of Nations*. New York: Knopf, 1991.

Roche, John P. "The Founding Fathers: A Reform Caucus in Action." In *The Formation of the Constitution*, 115–22. Ed. R. F. Jones. New York: R. E. Krieger, 1978.

Roloff, M., and Berger, C. R., eds. *Social Cognition and Communication*. Beverly Hills, Calif.: Sage, 1982.

Ross, Robert T., et al. *Lives of the Mentally Retarded*. Stanford, Calif.: Stanford University Press, 1985.

Rossi, Alice S. "A Biosocial Perspective on Parenting." *Daedalas* 106 (Spring 1977): 1–31.

——. "On the Reproduction of Mothering: A Methodological Debate." In *The Psychology of Women*, 265–73. Ed. Mary Ruth Walsh. New Haven, Conn.: Yale University Press, 1987.

Rossi, P., and Dentler, R. A. *The Politics of Urban Renewal*. Glencoe, Ill.: The Free Press, 1961.

Rossiter, Clinton. *1787: The Grand Convention*. New York: Mentor, 1966.

Rothbart, Mary K., and Posner, Michael I. "Temperament and the Development of Self-Regulation." In *The Neuropsychology of Individual Differences*, 93–123. Ed. L. C. Hartlage and C. F. Telzrow. New York: Plenum, 1985.
Rubenstein, Richard. *Alchemists of Revolution*. New York: Basic Books, 1987.
Ryan, Allan. *Quiet Neighbors*. New York: Harcourt, Brace, Jovanovich, 1984.
Sachs, Oliver. *The Man Who Mistook His Wife for a Hat*. New York: Summit, 1985.
Safire, William. *Before the Fall*. Garden City, N.Y.: Doubleday, 1975.
——. "The 109th Element." *New York Times*, January 17, 1985, 23.
Sanday, Peggy Reeves. *Female Power and Male Dominance*. Cambridge: Cambridge University Press, 1981.
Scarr, S., and McCartney, K. "How People Make Their Own Environments: A Theory of Genotype-Environment Effects." *Child Development* 54 (1983): 424–35.
Schlesinger, Arthur M., Jr. *The Politics of Upheaval*. Boston: Houghton Mifflin, 1960.
Schopler, E. "Introduction." In *Autism in Adolescents and Adults*. Ed. E. Shopler and G. Mesibov. New York: Plenum, 1983.
Schubert, Glendon. "Brain Science and Political Thinking." Paper presented at the meeting of the International Society of Political Psychology, Mannheim, Germany, 1981a.
——. "Some Implications of Brain Science for Political Science." Paper presented at the meeting of the Western Political Science Association, Denver, Colorado, 1981b.
——. *Evolutionary Politics*. Carbondale: Southern Illinois, 1989.
——. *Sexual Politics and Political Feminism*. Greenwich, Conn.: JAI Press, forthcoming.
Schumpeter, Joseph. *Imperialism and Social Classes*. New York: Kelley, 1951.
Scott, Peter B. *The Robotics Revolution*. Oxford, Eng.: Basil Blackwell, 1984.
Seager, J., and Olson, A. *Women in the World*. New York: Simon & Schuster, 1986.
Segal, Nancy. "Cooperation, Competition, and Altruism in Human Twinships: A Sociobiological Approach." In *Sociobiological Perspectives on Human Development*, 68–206. Ed. K. B. MacDonald. New York: Springer-Verlag, 1988.
Segev, Samuel. *The Iranian Triangle*. New York: The Free Press, 1988.
Simmel, Georg. *Conflict and the Web of Group Affiliations*. Trans. R. Bendix. New York: The Free Press, 1955.
Simpson, G. G. *The View of Life*. New York: Harcourt, Brace & World, 1964.
Sinnott, E. W. "The Biological Basis of Democracy." *Yale Review* 35 (1945): 61–73.
Smith, Henrik. *The Power Game*. New York: Random House, 1988.
Smith, M. Brewster, Bruner, Jerome, and White, Robert. *Opinions and Personality*. New York: Wiley, 1964.
Soames, Mary. *Clementine Churchill*. Boston: Houghton Mifflin, 1979.
Somit, A. *Biology and Politics*. Paris: Mouton, 1976.
Sorel, Nancy, ed. *Ever Since Eve*. New York: Oxford University Press, 1984.
Sorokin, P. A. *Social and Cultural Mobility*. Glencoe, Ill.: The Free Press, 1959.
Sowell, T. *The Economics and Politics of Race*. New York: Quill, 1983.
Sperry, R. W. *Science and Moral Priority*. New York: Columbia University Press, 1983.
Springer, S., and Deutsch, G. *Left Brain, Right Brain*. New York: W. H. Freeman, 1989.
Stenhouse, David. *The Evolution of Intelligence*. London: George Allen, 1973.
Sternberg, Robert J. "Cognitive Approaches to Intelligence." In *Handbook of Intelligence*, 59–118. Ed. Benjamin Wolman. New York: Wiley, 1985.
——. "Inside Intelligence." *American Scientist* 74 (1986): 137–43.
——. *The Triarchic Mind*. New York: Viking Penguin, 1988.
Stevens, Mark. *The Insiders*. New York: G. P. Putnam's Sons, 1987.

Stouffer, Samuel. *Communism, Conformity, and Civil Liberties*. New York: Doubleday, 1958.
Tennov, Dorothy. *Love and Limerance: The Experience of Being in Love*. Chelsea, Mich.: Scarborough House, 1989.
Thompson, Hunter S. *Hell's Angels*. New York: Ballantine, 1967.
Tiger, L. *Men in Groups*. New York: Random House, 1969.
Tiger, Lionel, and Shepher, Joseph. *Women in the Kibbutz*. New York: Harvest, 1975.
Tocqueville, Alexis de. *Democracy in America*. New York: Vintage, 1955.
Toffler, Alvin. *Future Shock*. New York: Random House, 1970.
Tower Commission Report. New York: Bantam, 1987.
Trasler, Gordon. "Biogenetic Factors." In *Handbook of Juvenile Delinquency*, 184–215. Ed. H. C. Quay. New York: Wiley, 1987.
Treverton, Gregory. *Covert Action*. New York: Basic Books, 1987.
Trivers, R. "Sociobiology and Politics." In *Sociobiology and Human Politics*. Ed. E. White. Lexington, Mass.: Lexington Books, 1981.
Vandenberg, Steven G., and Vogler, G. F. "Genetic Determinants of Intelligence." In *Handbook of Intelligence*, 3–59. Ed. B. B. Wolman. New York: Wiley, 1985.
Van Doren, Carl. *Benjamin Franklin*. New York: Viking, 1938.
Veevers, J. E. "Voluntary Childlessness: A Neglected Area of Family Study." In *Pronatalism*. Ed. E. Peck and J. Senderowitz. New York: Crowell, 1974.
Volkman, Ernest, and Buggett, Blaine. *Secret Intelligence*. New York: Doubleday, 1989.
Vroom, Victor. *Work and Motivation*. New York: Wiley, 1967.
Waddington, C. H. *The Ethical Animal*. New York: Atheneum, 1961.
Walker, Alice. "In Search of Our Mothers' Gardens." In *Working It Out,* 92–102. Ed. Sara Ruddick and Pamela Daniels. New York: Pantheon, 1977.
Walsh, Mary, ed. *The Psychology of Women*. New Haven, Conn.: Yale University Press, 1987.
Watts, M., ed. *Biopolitics and Gender*. New York: Haworth, 1983.
Webber, M. M. "Culture, Territoriality, and the Elastic Mile." *Papers and Proceedings of the Regional Science Association* 13 (1964): 59–69.
Weinstein, Allen. *Perjury: The Hiss-Chambers Case*. New York: Knopf, 1978.
Weiss, M. *The Clustering of America*. New York: Harper, 1988.
Weissman, Alden E., and Gorman, B. S. "The Emergence of Human Awareness and Concepts of Time." In *The Personal Experience of Time*, 3–55. Ed. B. S. Gorman and A. E. Weissman. New York: Plenum, 1977.
White, E. *The Conversion of Wants into Demands*. Master's Thesis. Department of Political Science, University of Chicago, 1959.
——. "Genetic Diversity and Political Life: Toward a Populational Interaction Paradigm." *Journal of Politics* 34 (1972): 1203–42.
——. "The Neurobiological Basis of Human Action." Paper presented at Western Political Science Association meeting, Denver, Colorado, 1981a.
——. "Sociobiology, Neurobiology, and Political Socialization." *Micropolitics* 1 (1981b): 113–44.
——. "Brain Science and the Emergence of Neuropolitics." *Politics and the Life Sciences* I (1982a).
——. "Clouds, Clocks, Brains, and Political Learning." *Micropolitics* 2 (1982b): 279–309.
——. "Brains, Bonds, and Bureaucracy." In *Biology and Bureaucracy*. Ed. E. White and J. Losco. Lanham, Md.: University Press of America, 1986.

——. "Self-Selection and Political Life: The Neuropolitics of Alienation—The Trapped and The Overwhelmed." *Politics and the Life Sciences* 17, no. 2 (1989): 154–85.

——. *The End of the Empty Organism: Neurobiology and the Sciences of Human Action*. Westport, Conn.: Praeger, 1992.

White, T. H. *Breach of Faith*. New York: Atheneum, 1975.

Wiegele, T. *Biopolitics*. Boulder, Colo.: Westview, 1979.

——. "The Emerging Significance of Biotechnology for the Study of International Relations." Paper prepared for the annual convention of the Association of Politics and the Life Sciences and the American Political Science Association, August 30–September 2, 1990, San Francisco, California.

Willhoite, R. "Political Evolution and Legitimacy: The Biocultural Origins of Hierarchical Organizations." In *Biology and Bureaucracy*. Ed. E. White and J. Losco. Lanham, Md.: University Press of America, 1986.

Williams, Juanita. *Psychology of Women*. New York: Norton, 1977.

Wilson, E. O. *On Human Nature*. Cambridge, Mass.: Harvard University Press, 1978.

Wilson, James Q. "Planning and Politics: Citizen Participation in Urban Renewal." In *Urban Renewal: People, Politics, and Planning*, 287–301. Ed. J. Bellson and M. Harsknecht. Garden City, N.Y.: Anchor, 1967.

——. *American Government*. Lexington, Mass.: D. C. Heath, 1983.

Wilson, J. Q., and Herrnstein, R. J. *Crime and Human Nature*. New York: Simon & Schuster, 1985.

Witkin-Lanoil, Georgia. *The Female Stress Syndrome*. New York: Newmarket Press, 1984.

Wolf, D. G. *The Lesbian Community*. Berkeley: University of California Press, 1979.

Wolferen, Karel van. *The Enigma of Japanese Power*. New York: Knopf, 1989.

Wood, Gordon. *The Creation of the American Republic 1776–1787*. Chapel Hill: University of North Carolina Press, 1969.

——. "Interests and Disinterestness in the Making of the Constitution." In *Beyond Confederation*. Ed. R. Beeman, S. Botein, and E. C. Carter. Chapel Hill: University of North Carolina Press, 1987.

Woodward, Bob. *Veil*. New York: Simon & Schuster, 1987.

Wright, Esmond. *Franklin of Philadelphia*. Cambridge, Mass.: The Belknap Press, 1986.

Yeo-chi King, Ambrose. "Kuan-hsi and Network Building: A Sociological Interpretation." *Daedalus* 120 (1991): 63–84.

Zaleznik, Abraham. "Managers and Leaders: Are They Different?" In *Harvard Business Review: On Human Relations*, 162–79. New York: Harper & Row, 1979.

Zuckerman, M. "A Biological Theory of Sensation-Seeking." In *Biological Bases of Sensation-Seeking, Impulsivity and Anxiety*. Ed. M. Zuckerman. Hillsdale, N.J.: Erlbaum, 1983.

Index

About the Author

ELLIOTT WHITE is Professor of Political Science at Temple University, Japan. He is the author of *The End of the Empty Organism* (Praeger, 1992), the editor of *Sociobiology and Human Politics* (1981), and the co-editor (with Joseph Losco) of *Biology and Bureaucracy* (1986).